Basics for Builders:

FRAMING & ROUGH CARPENTRY

Scot Simpson

Illustrated by Gary Burhop

Basics for Builders:

FRAMING & ROUGH CARPENTRY

Scot Simpson

Copyright 1991

R.S. Means Company, Inc.

Construction Publishers & Consultants
Construction Plaza
63 Smiths Lane
Kingston, MA 02364-0800
(781) 585-7880

The editors for this book were Howard Chandler, Mary P. Greene and Timothy Jumper; the production supervisor was Marion E. Schofield. Typesetting was supervised by Joan Marshman. The book and cover were designed by Norman R. Forgit. Editor Reference Books: Mary Greene.

Printed in the United States of America

10

Library of Congress Catalog Number 91-192153

ISBN 0-87629-251-1

TABLE OF CONTENTS

Chapter 5: Roof Framing

Chapter 6: Doors, Windows, and Stairs

Chapter 7: Layout

INTRODUCTION

A few years ago, I sought a textbook to use in teaching my framer trainees the basics. I felt they should all learn the same framing method to ensure that they would correctly interpret my instructions, and execute them with predictable and consistent results. I wanted the text to be easy enough so that they could read and absorb its contents quickly. Although many carpentry books were available, I was unable to find one that met these requirements.

Combining this need for a text with my 21 years experience framing in many states across the country (including 48 apartment projects, 9 condominium projects, 8 office buildings, 6 hotels, and 27 houses), and having gone to school evenings to acquire a Masters degree in Business Administration, I decided to write my own training manual that would be a clear, concise, easy-to-use outline of modern methods of framing, covering all aspects of the trade—from definitions of framing terms to guidelines for the management of a framing crew.

This manual was developed for framing contractors and framer trainees, but can be used equally well by many others, including general carpenters, renovators, remodelers, handymen, and do-it-yourself weekenders; in short, anyone interested in learning the basic skills of framing.

Every framer pounding nails today will have techniques that he prefers to some of those described here. There are also some tricks of the trade that can only be learned from old-timers on the job, and it's a wise apprentice who keeps his eyes and ears open for them. What I have tried to do in this manual is bring together the best techniques from different parts of the country, setting a standard for my trainees and crews.

My father, who was a professional journalist, told me years ago that a picture is worth 10,000 words. 10,000 written words represents a lot of time and often frustration for many framers. For this reason, I have used photographs and drawings extensively throughout the book to illustrate as much of the text as possible.

Another instructional technique that you will find in this manual is the step-by-step procedure for various tasks, for example, the sixteen-step wall framing sequence (W1–W16). These steps are listed in a boxed table of contents on the first page of each chapter. This step-by-step method standardizes the framing process so that it is easy to learn. This is especially important in framing because framers are often involved in different tasks from day-to-day, and a month or more may go by before a particular task is repeated. The step-by-step sequence method makes it easy to refresh your memory on the procedure to follow for a particular task.

This manual is designed to be "Framer Friendly." I hope that you find it friendly to you.

SCOT SIMPSON

Chapter One

THE TRADE OF WOOD FRAMING

The trade of wood framing is comprised of the rough carpentry skills needed to produce the "skeleton" of a building and its first layer of "skin." The skeleton consists of the structural lumber forming the floors, walls, and roof; the skin consists of the lumber which encloses the skeleton, and provides a surface for subsequent layers of protective and decorative finish materials.

This chapter is an illustrated review of the most basic tools, materials, and terminology of the framer. This information is so basic that it usually is not even taught on the job site, so if you don't know it when you arrive for work, you will have to play a guessing game or ask a lot of questions.

The illustrations with terms serve as a handy reference, and help to reduce confusion when different words are used for the same item. This happens a lot because framers move from job site to job site, and work with different people. For example, bottom plates are often known as *sole plates*, backers as *partitions*, and trimmers as *cripples*. It doesn't matter what they are called as long as you know what they are. There is also a more detailed list of framing terms with definitions at the back of the book.

The suggested organization for a framing tool truck presented in this chapter is just an example of how a truck could be set up for tool storage. Its purpose is, once again, to reduce confusion and make the job easier. It is amazing how much time can be spent looking for tools and nails if they aren't put where a framer expects them to be.

FRAMING TERMS

Bearing Walls

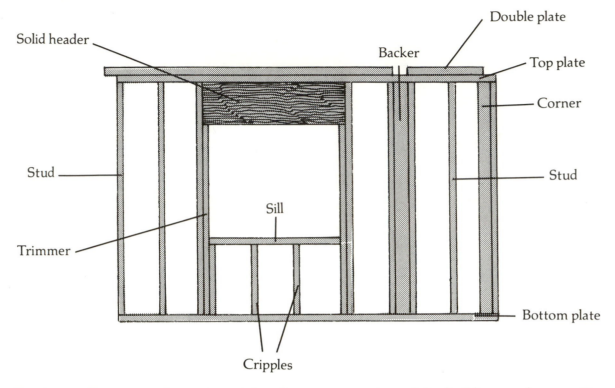

Bearing walls support the main weight of an upper portion of the building, such as a ceiling, floor, or roof. Nonbearing walls provide little or no support to those upper portions. Remove nonbearing walls and the upper portions will stand; remove bearing walls and the upper portions will fall.

Nonbearing Walls

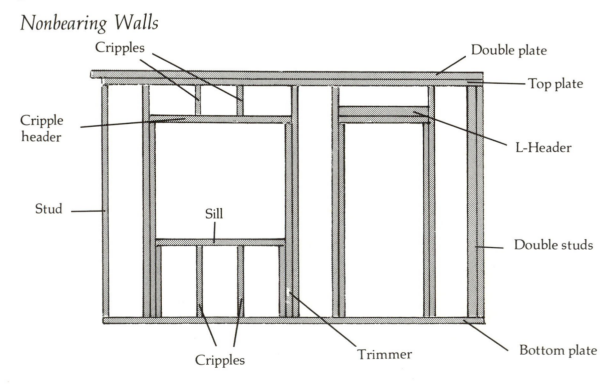

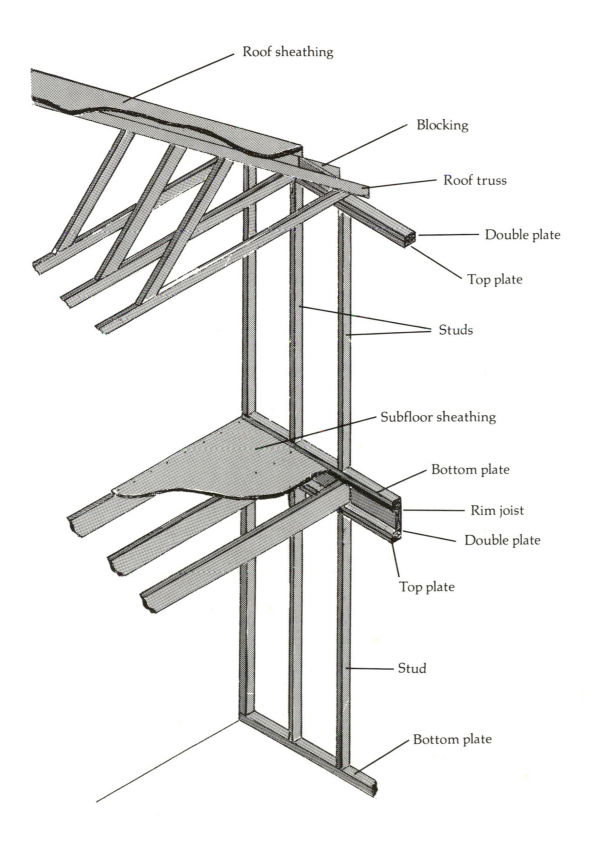

Roof sheathing

Blocking

Roof truss

Double plate

Top plate

Studs

Subfloor sheathing

Bottom plate

Rim joist

Double plate

Top plate

Stud

Bottom plate

FRAMING LUMBER

Lumber is sized in "nominal," as opposed to "actual," dimensions. A nominal dimension rounds off the actual dimension to the next highest whole number. For example, a piece of lumber which actually measures 1½" × 3½" is rounded off to the nominal 2" × 4".

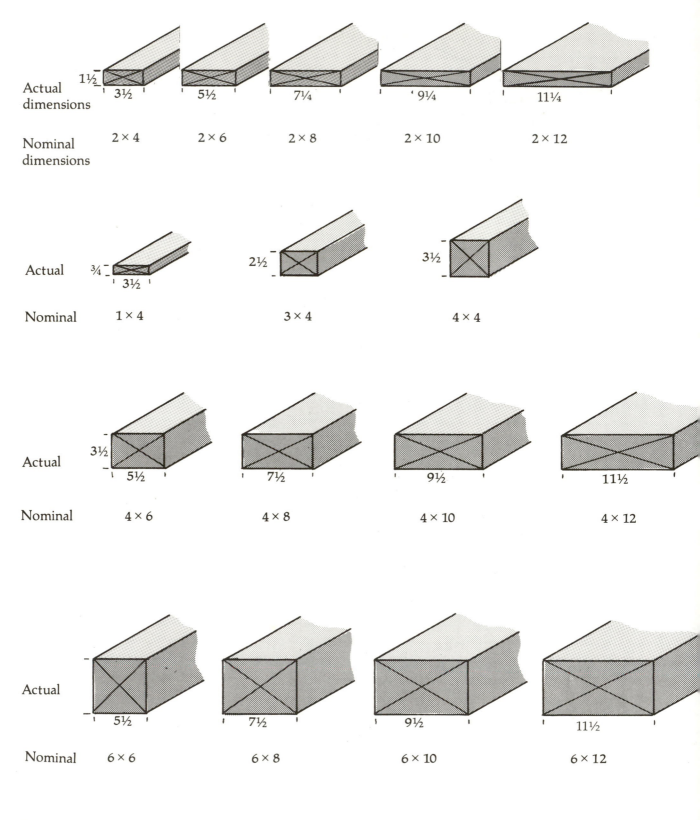

Actual dimensions	1½ 3½	5½	7¼	9¼ 11¼
Nominal dimensions	2 × 4	2 × 6	2 × 8	2 × 10 2 × 12

Actual	¾ 3½	2½	3½
Nominal	1 × 4	3 × 4	4 × 4

Actual	3½ 5½	7½	9½	11½
Nominal	4 × 6	4 × 8	4 × 10	4 × 12

Actual	5½	7½	9½	11½
Nominal	6 × 6	6 × 8	6 × 10	6 × 12

FRAMING SHEATHING

Plywood

½″ sheet
⅝″ sheet
¾″ sheet

Plywood comes in 4′ × 8′ sheets. The thicknesses most commonly used in framing are ½″, ⅝″, and ¾″.

T&G Plywood (tongue and groove)

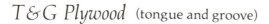

GWB

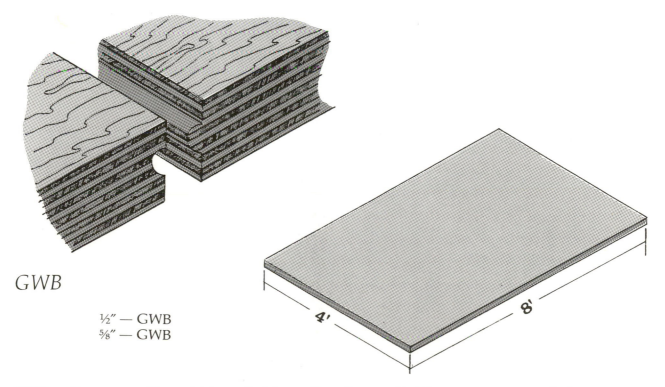

½″ — GWB
⅝″ — GWB

GWB = Gypsum wallboard (also called drywall or sheetrock). The most common thicknesses are ½″ and ⅝″.

LUMBER AND PLYWOOD GRADES

Lumber and plywood are graded for strength and different uses. Each piece of lumber is stamped for identification before it is shipped. Architects specify grades of lumber and plywood for various purposes, and framers need to make sure the right wood is used.

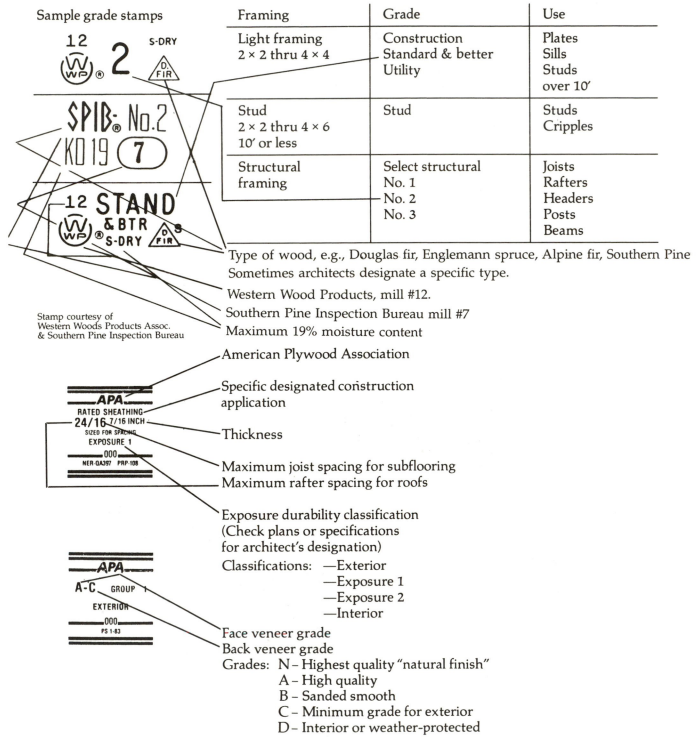

Sample grade stamps

Framing	Grade	Use
Light framing 2 × 2 thru 4 × 4	Construction Standard & better Utility	Plates Sills Studs over 10′
Stud 2 × 2 thru 4 × 6 10′ or less	Stud	Studs Cripples
Structural framing	Select structural No. 1 No. 2 No. 3	Joists Rafters Headers Posts Beams

Type of wood, e.g., Douglas fir, Englemann spruce, Alpine fir, Southern Pine Sometimes architects designate a specific type.

Western Wood Products, mill #12.

Southern Pine Inspection Bureau mill #7

Maximum 19% moisture content

Stamp courtesy of
Western Woods Products Assoc.
& Southern Pine Inspection Bureau

American Plywood Association

Specific designated construction application

Thickness

Maximum joist spacing for subflooring
Maximum rafter spacing for roofs

Exposure durability classification
(Check plans or specifications
for architect's designation)

Classifications: —Exterior
                    —Exposure 1
                    —Exposure 2
                    —Interior

Face veneer grade
Back veneer grade
Grades: N – Highest quality "natural finish"
          A – High quality
          B – Sanded smooth
          C – Minimum grade for exterior
          D – Interior or weather-protected

(Stamp, courtesy of American Plywood Association)

6

FRAMING NAILS

Frequently Used

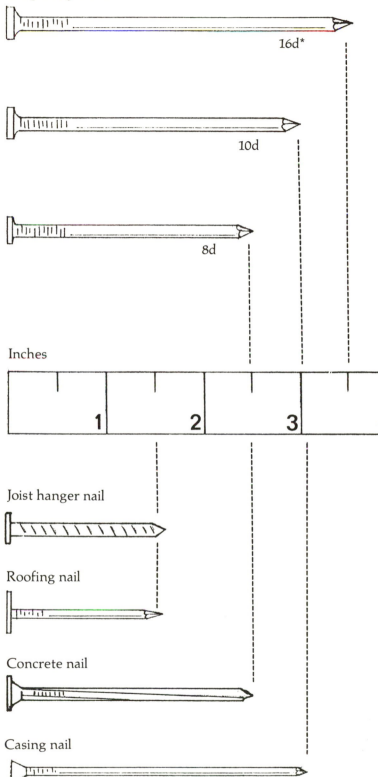

16d*

10d

8d

Inches

| | | 1 | | 2 | | 3 | | 4 |

Joist hanger nail

Roofing nail

Concrete nail

Casing nail

*d = Penny. The abbreviation comes from the Roman word "denarius" meaning coin, which the English adapted to penny. It originally referred to the cost of a specific nail per 100. Today it refers only to nail size.

FRAMING TOOLS

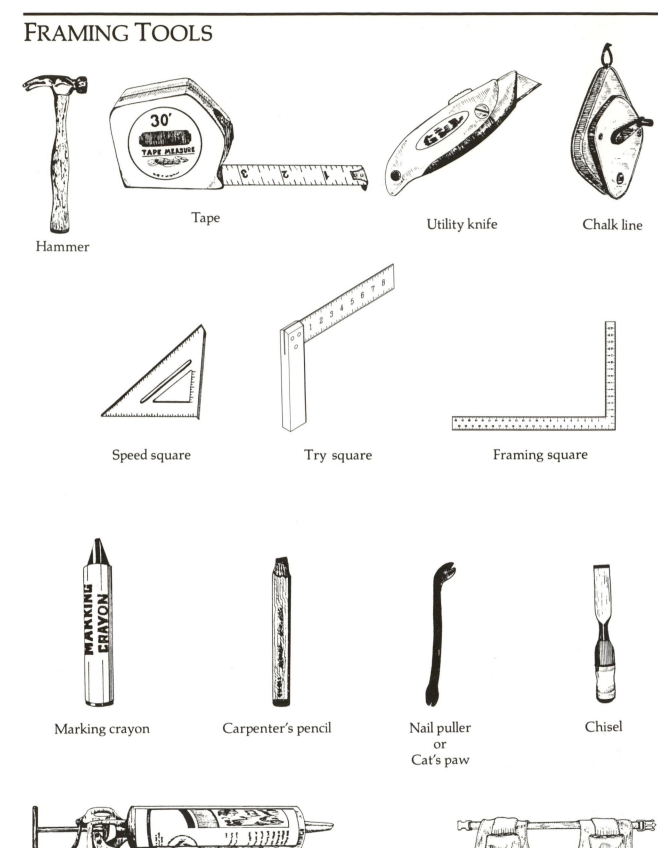

Hammer

Tape

Utility knife

Chalk line

Speed square

Try square

Framing square

Marking crayon

Carpenter's pencil

Nail puller
or
Cat's paw

Chisel

Glue gun

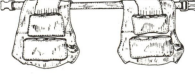

Tool pouch

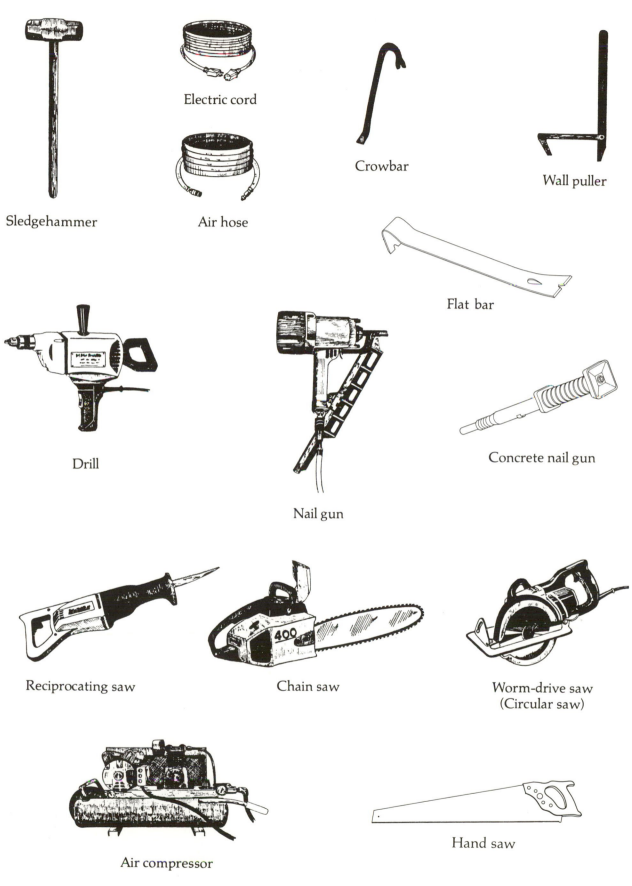

Sledgehammer

Electric cord

Air hose

Crowbar

Wall puller

Flat bar

Drill

Nail gun

Concrete nail gun

Reciprocating saw

Chain saw

Worm-drive saw
(Circular saw)

Air compressor

Hand saw

FRAMING TOOL TRUCK

Typical Layout for a 14' Step Van

Organizing your tools helps keep them in good condition and helps you find them when you need them, thereby saving valuable time on the job.

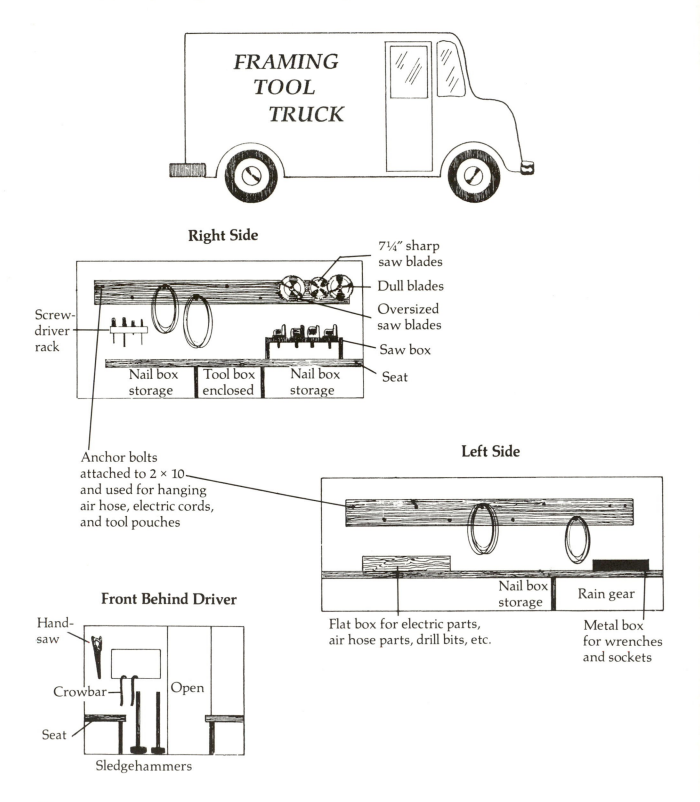

FRAMING
TOOL
TRUCK

Right Side

7¼" sharp saw blades

Dull blades

Oversized saw blades

Saw box

Seat

Screwdriver rack

Nail box storage | Tool box enclosed | Nail box storage

Anchor bolts attached to 2 × 10 and used for hanging air hose, electric cords, and tool pouches

Left Side

Nail box storage

Rain gear

Flat box for electric parts, air hose parts, drill bits, etc.

Metal box for wrenches and sockets

Front Behind Driver

Handsaw

Crowbar

Open

Seat

Sledgehammers

CUTTING LUMBER

Accuracy in measuring, marking, and cutting lumber is a very important framing skill to master.

Periodic checks should be made of the condition and accuracy of tape measures and the squareness of saw tables and blades.

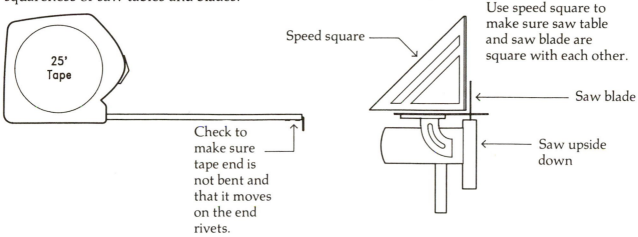

25'
Tape

Check to make sure tape end is not bent and that it moves on the end rivets.

Speed square

Use speed square to make sure saw table and saw blade are square with each other.

Saw blade

Saw upside down

A typical saw blade removes a channel of wood approximately ⅛" wide, called a "kerf." This must be taken into consideration when you make a cut.

Suppose you want to cut a board 25" long. Measure and make a mark at 25", then square a line through the mark with a try-square. The "work piece" — the 25" piece you want to use — will be to the left of the line; the "waste piece" will be to the right. Guide your saw along the right edge of the line so the kerf is made in the waste piece. If your cut is perfectly made, the work piece will be left showing exactly half the width of your pencil line, and will measure exactly 25". Thus, the old carpenter's saying: "Leave the line."

EX: Cut a 25" piece

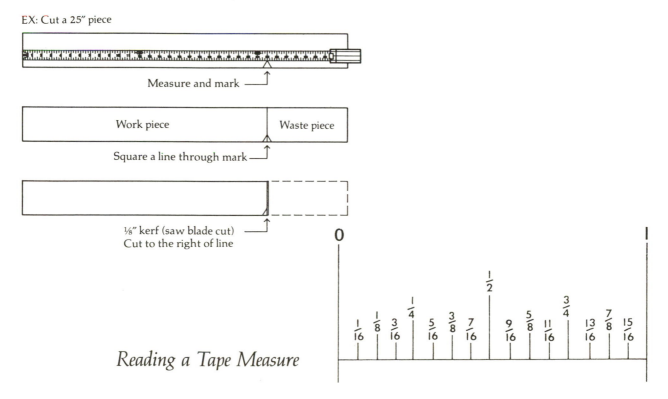

Measure and mark

Work piece | Waste piece

Square a line through mark

⅛" kerf (saw blade cut)
Cut to the right of line

Reading a Tape Measure

Chapter Two

NAILING PATTERNS

Building codes and generally accepted practices were followed in developing these nailing patterns, which apply when the plans do not specify anything different. When the plans do call for other nailing patterns, be sure to follow them.

If you are framing every day, these patterns will soon become second nature. For the part-time framer, this chapter can serve as a quick reference.

Contents

NAIL TOP PLATE TO STUDS

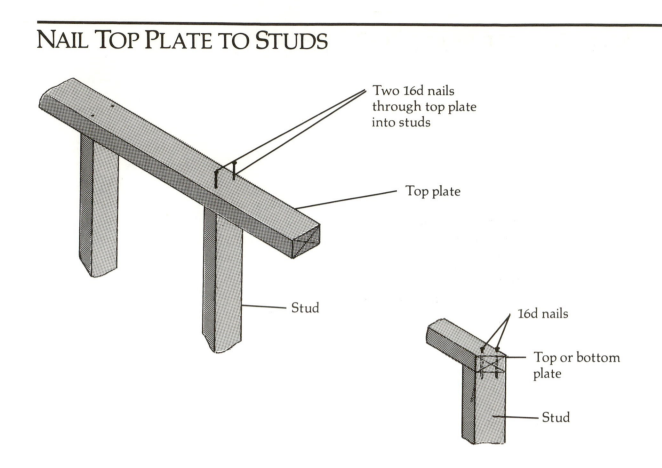

Two 16d nails through top plate into studs

Top plate

Stud

16d nails

Top or bottom plate

Stud

NAIL BOTTOM PLATE TO STUDS

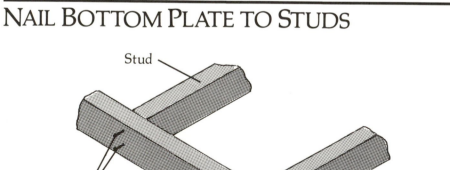

Stud

Two 16d nails through bottom plate into studs

Bottom plate

NAIL DOUBLE PLATE TO TOP PLATE

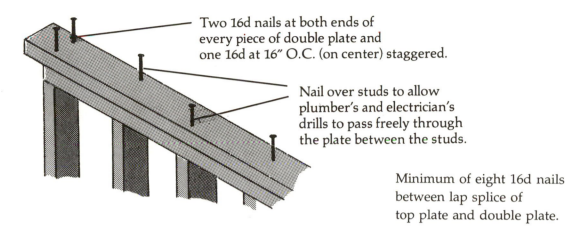

Two 16d nails at both ends of every piece of double plate and one 16d at 16" O.C. (on center) staggered.

Nail over studs to allow plumber's and electrician's drills to pass freely through the plate between the studs.

Minimum of eight 16d nails between lap splice of top plate and double plate.

NAIL CORNER

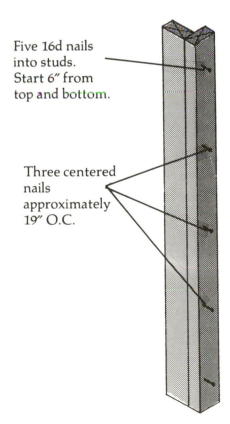

Five 16d nails into studs. Start 6" from top and bottom.

Three centered nails approximately 19" O.C.

NAIL WALLS TOGETHER OR NAIL DOUBLE STUDS

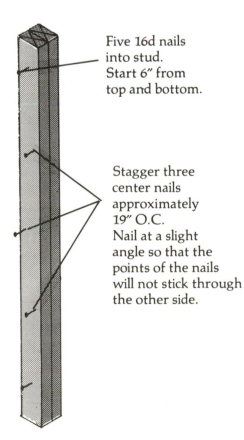

Five 16d nails into stud. Start 6" from top and bottom.

Stagger three center nails approximately 19" O.C. Nail at a slight angle so that the points of the nails will not stick through the other side.

NAIL TRIMMER TO STUD

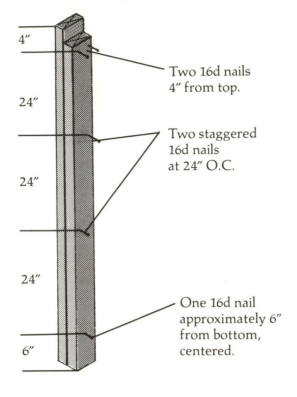

4"

24"

24"

24"

6"

Two 16d nails
4" from top.

Two staggered
16d nails
at 24" O.C.

One 16d nail
approximately 6"
from bottom,
centered.

CONCRETE NAILING

Bearing Walls

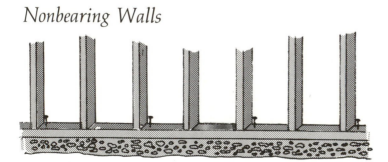

One 2½" concrete nail at 16" O.C.,
or at every stud.

Nonbearing Walls

One 2½" concrete nail at 32" O.C.,
or at every other stud.

Nail Bearing and Nonbearing Walls to Floor Perpendicular to Joists

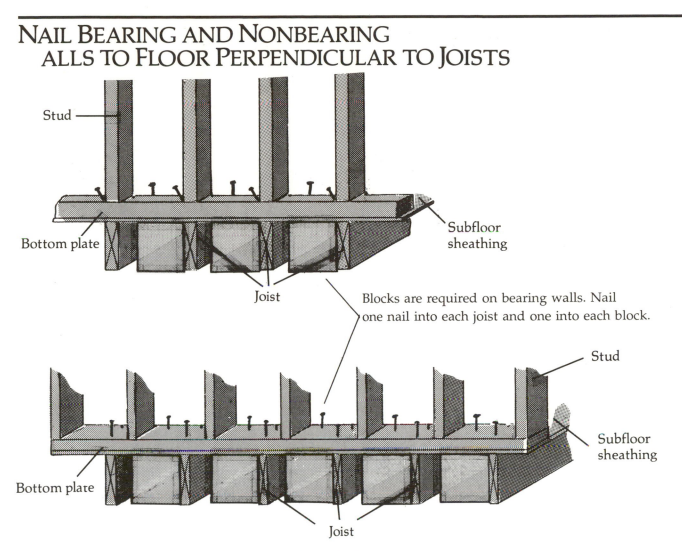

Stud

Bottom plate

Subfloor sheathing

Joist

Blocks are required on bearing walls. Nail one nail into each joist and one into each block.

Stud

Subfloor sheathing

Bottom plate

Joist

Nail nonbearing walls similarly, using one nail instead of two.
Braced wall panels and shear walls require three 16d nails at 16" O.C.

Nail Bearing and Nonbearing Walls to Floor Parallel to Joists

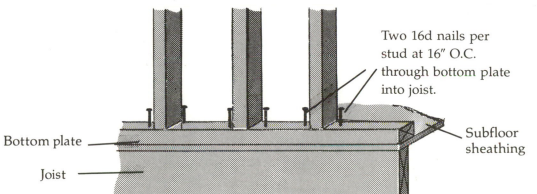

Two 16d nails per stud at 16" O.C. through bottom plate into joist.

Bottom plate

Subfloor sheathing

Joist

Nail nonbearing walls similarly, using one nail instead of two.
Braced wall panels and shear walls require three 16d nails at 16" O.C.

NAIL HEADER TO STUD

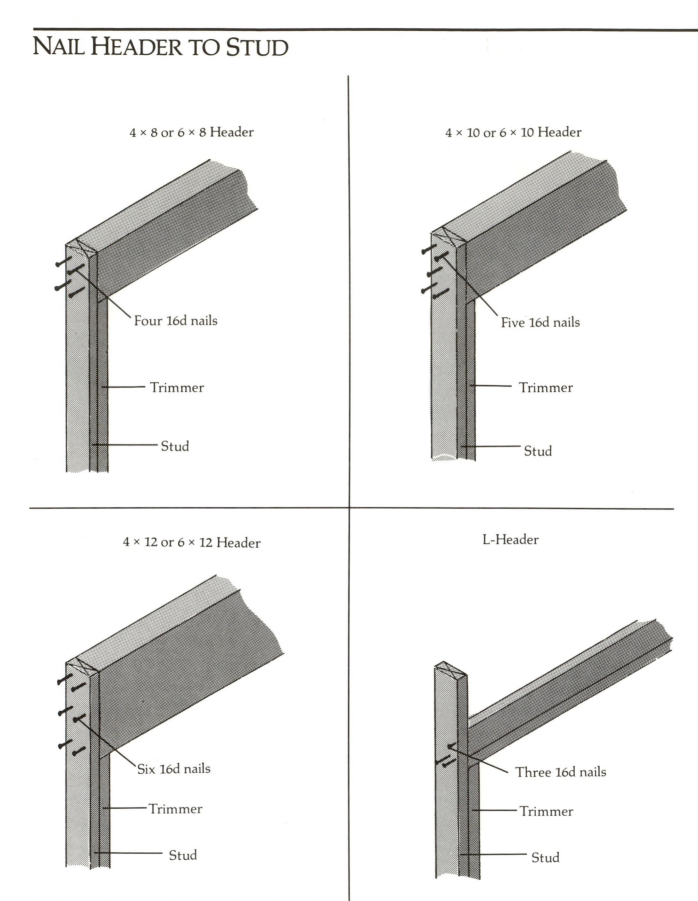

4 × 8 or 6 × 8 Header

Four 16d nails

Trimmer

Stud

4 × 10 or 6 × 10 Header

Five 16d nails

Trimmer

Stud

4 × 12 or 6 × 12 Header

Six 16d nails

Trimmer

Stud

L-Header

Three 16d nails

Trimmer

Stud

Headers made up of 2× lumber with ½" plywood sandwiched between should be nailed similarly.

NAIL LET-IN BRACING

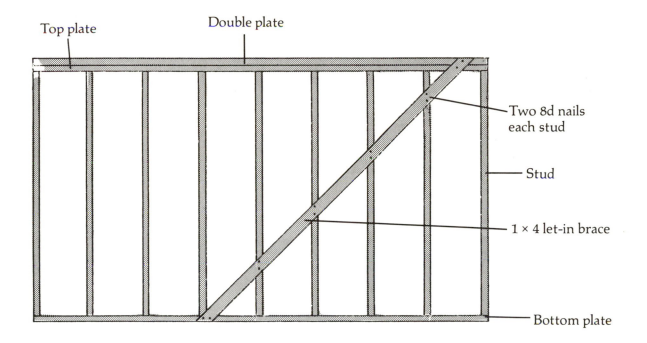

Top plate

Double plate

Two 8d nails each stud

Stud

1 × 4 let-in brace

Bottom plate

- Two 8d nails at stud intersections
- 45 degree is best angle
- Cross five studs if possible

NAIL END OF JOIST

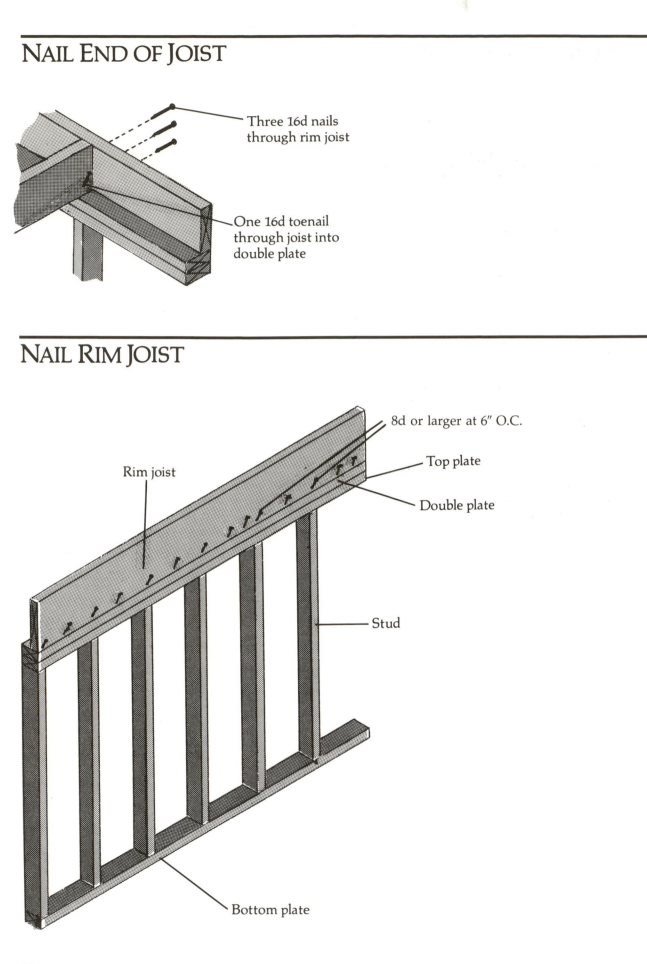

Three 16d nails through rim joist

One 16d toenail through joist into double plate

NAIL RIM JOIST

8d or larger at 6″ O.C.

Top plate

Double plate

Rim joist

Stud

Bottom plate

NAIL SUBFLOOR SHEATHING

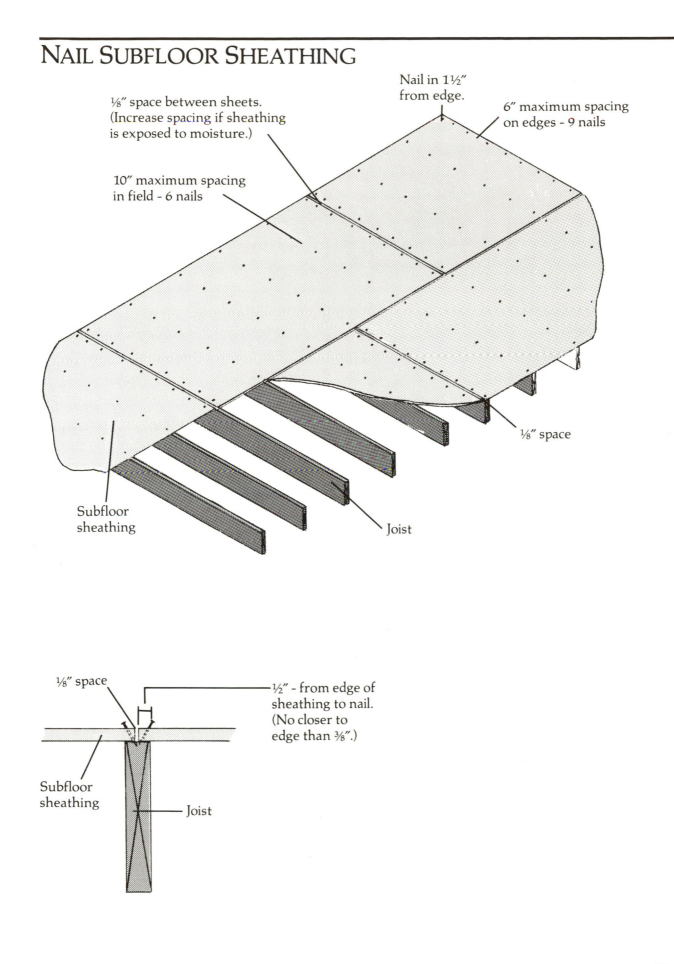

⅛" space between sheets.
(Increase spacing if sheathing
is exposed to moisture.)

Nail in 1½"
from edge.

6" maximum spacing
on edges - 9 nails

10" maximum spacing
in field - 6 nails

⅛" space

Subfloor
sheathing

Joist

⅛" space

½" - from edge of
sheathing to nail.
(No closer to
edge than ⅜".)

Subfloor
sheathing

Joist

Nail Header Joists and Multiple Joists

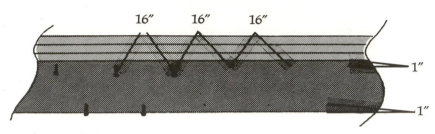

16"   16"   16"

1"

1"

16d at 16" O.C.
above studs

Nail Joist Blocking

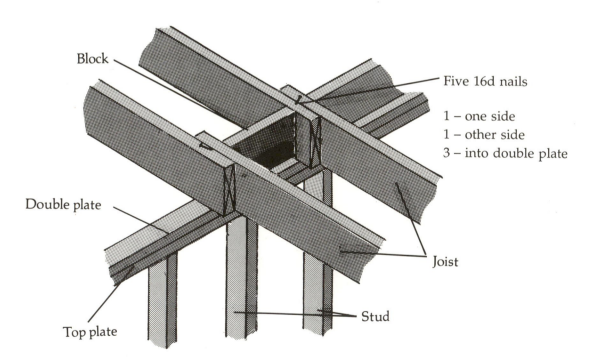

Block

Five 16d nails

1 – one side
1 – other side
3 – into double plate

Double plate

Joist

Top plate

Stud

NAIL LAPPING JOISTS

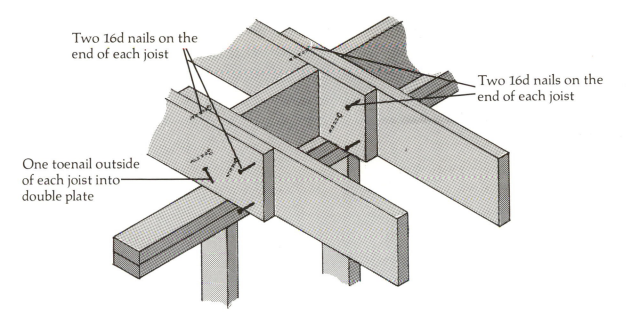

Two 16d nails on the end of each joist

Two 16d nails on the end of each joist

One toenail outside of each joist into double plate

- Six nails each lap
- Two 16d nails on the end of each joist
- One 16d nail from each joist down into double plate

NAIL DRYWALL BACKING

16d nail at 16" O.C. over each stud

Rim joist

Drywall

Double plate

2 × 4 backing

Top plate

Stud

Joist

Drywall backing

Top plate

Double plate

Stud

Drywall

NAIL TRUSSES TO WALL

Three 8d or larger nails toenailed

Roof truss

Stud

Double plate

Top plate

NAIL CEILING JOIST

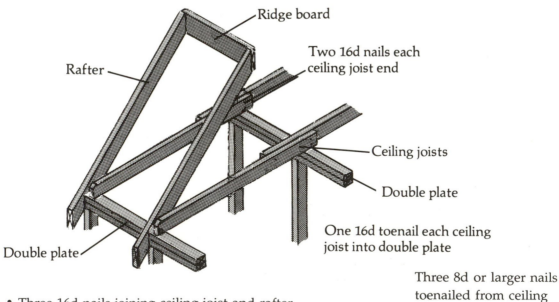

Ridge board

Rafter

Two 16d nails each ceiling joist end

Ceiling joists

Double plate

One 16d toenail each ceiling joist into double plate

Double plate

Three 8d or larger nails toenailed from ceiling joist into double plate

- Three 16d nails joining ceiling joist and rafter
- Three 8d or larger toenail from ceiling joist into double plate
- Three 8d or larger toenail from rafter into double plate

NAIL RAFTERS TO WALL

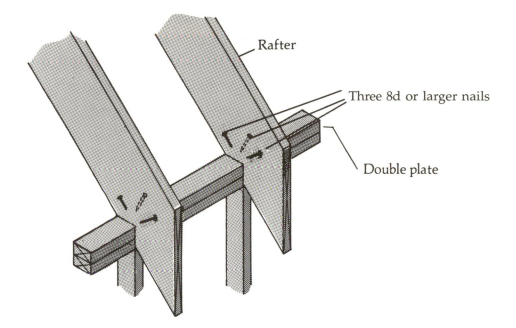

Rafter

Three 8d or larger nails

Double plate

NAIL BLOCKS

Block position
depends on exterior
finish and venting

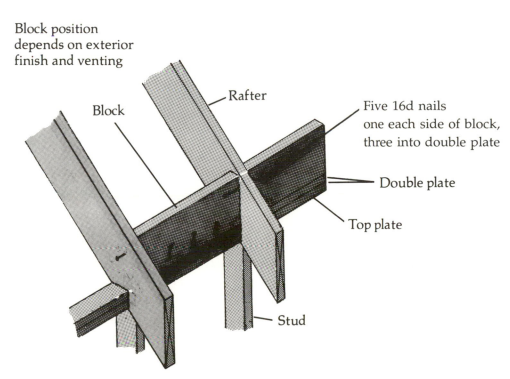

Block

Rafter

Five 16d nails
one each side of block,
three into double plate

Double plate

Top plate

Stud

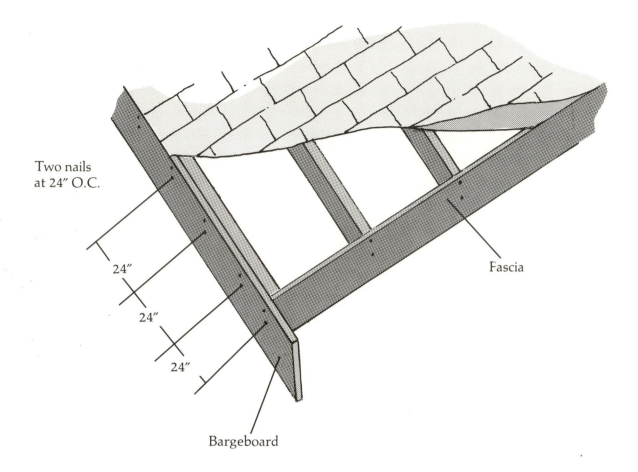

Two nails
at 24" O.C.

24"

24"

24"

Fascia

Bargeboard

- 1" fascia - use 8d galvanized box nails
- 2" fascia - use 16d galvanized box nails

Chapter Three

FLOOR FRAMING

This chapter illustrates the basic sequence for floor framing. Straight cuts and tight nailing make for a neat and professional job. Pay particular attention to the corners. It is important that they stay square and plumb up from the walls below, so the building does not gain or lose in size. Also pay close attention to laying the first sheet of subfloor sheathing. If it is laid straight and square, the entire subfloor will go down easily and require no unnecessary cuts. If you make a sloppy start on the first sheet, you'll struggle to make each sheet fit, you'll waste valuable time, and you won't be proud of the results.

F1- CROWN JOISTS AND PUT IN PLACE

Look for crown

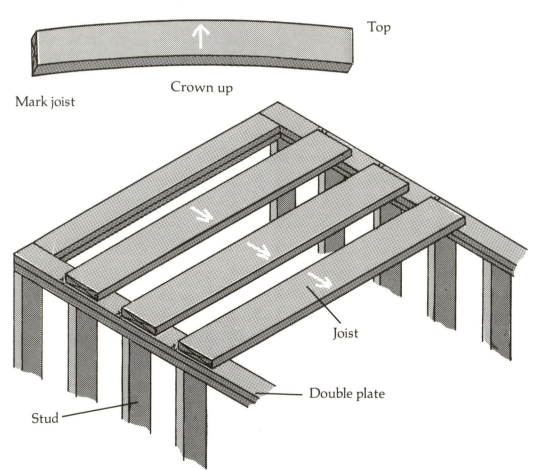

Top

Crown up

Mark joist

Joist

Double plate

Stud

Spread joists so crowns are in the same direction.

The crown is the highest point of a curved piece of lumber.

If the joists are sitting on a foundation instead of a stud wall, then a sill plate, or mudsill would be attached to the foundation, and the joists would sit on the plate or sill.

F2-Nail Rim Joists In Place and
F3-Cut Joists to Length

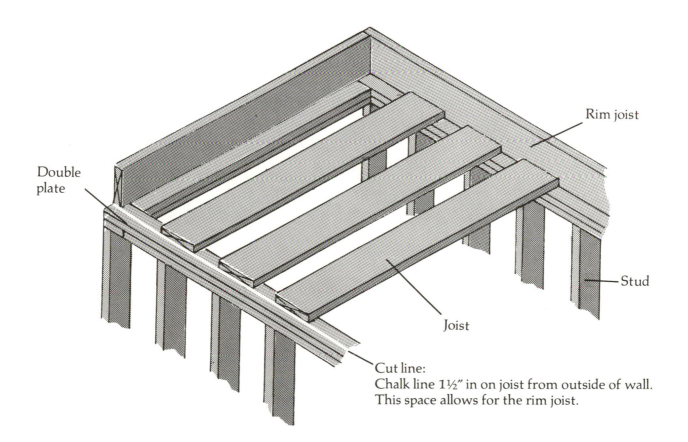

Double plate

Rim joist

Stud

Joist

Cut line:
Chalk line 1½" in on joist from outside of wall.
This space allows for the rim joist.

If joists lap over an interior wall, they can be rough-cut approximately two inches beyond the wall. Do not let lapped joists go more than six inches beyond the wall.

F4-Nail Joists in Place

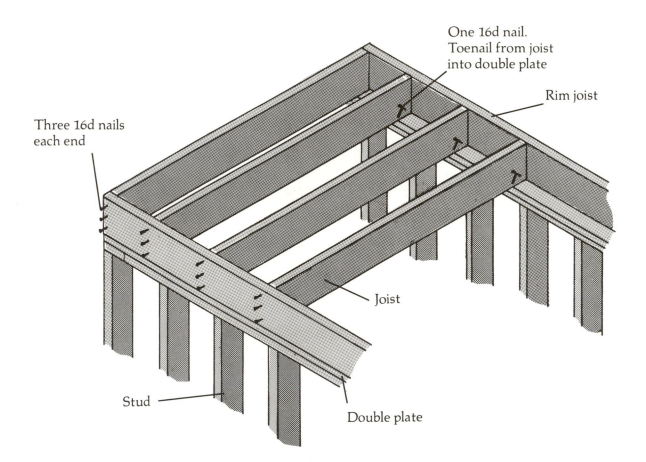

One 16d nail.
Toenail from joist
into double plate

Rim joist

Three 16d nails
each end

Joist

Stud

Double plate

Turn joists crown up and nail into place.

F5-FRAME OPENINGS IN JOISTS

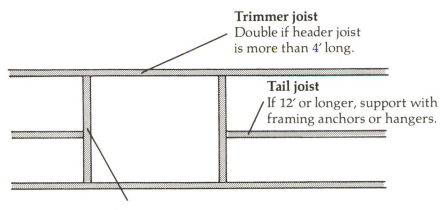

Trimmer joist
Double if header joist
is more than 4' long.

Tail joist
If 12' or longer, support with
framing anchors or hangers.

Header joist
Double if longer than 4'.
If longer than 6', support with
framing anchors or hangers.

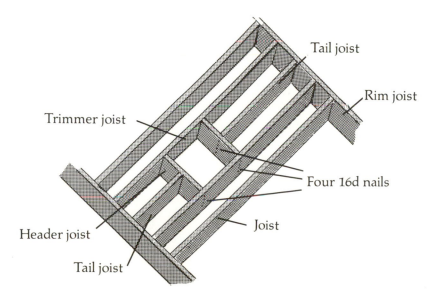

Tail joist

Rim joist

Trimmer joist

Four 16d nails

Header joist

Joist

Tail joist

F6-BLOCK BEARING WALLS
F7-NAIL JOISTS TO WALLS

Nail joists to walls to hold interior walls in place.

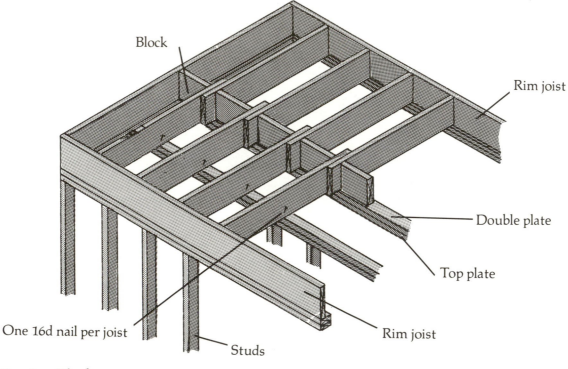

Block

Rim joist

Double plate

Top plate

One 16d nail per joist

Studs

Rim joist

Cutting Blocks

16" O.C. spacing
   Single joist blocks — $14^{7}/_{16}$"
   Lapped joist blocks — $12^{7}/_{8}$"

24" O.C. spacing
   Single joist blocks — $22^{7}/_{16}$"
   Lapped joist blocks — $20^{7}/_{8}$"

F8-Drywall Backing

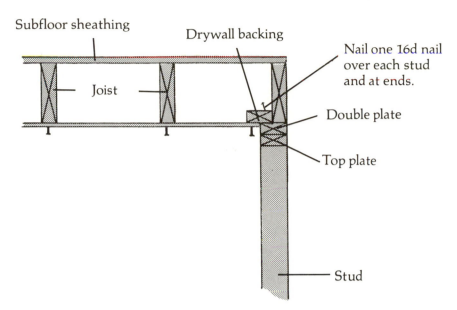

Subfloor sheathing

Drywall backing

Nail one 16d nail over each stud and at ends.

Joist

Double plate

Top plate

Stud

Whenever the distance from the edge of the wall to the joist is greater than 6", place drywall backing on top of the wall.

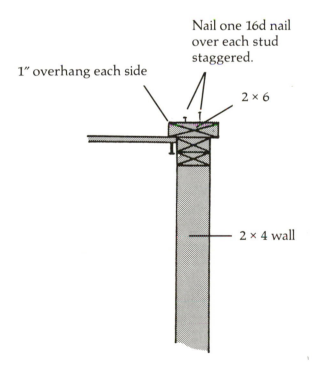

Nail one 16d nail over each stud staggered.

1" overhang each side

2 × 6

2 × 4 wall

F9-SUBFLOOR SHEATHING

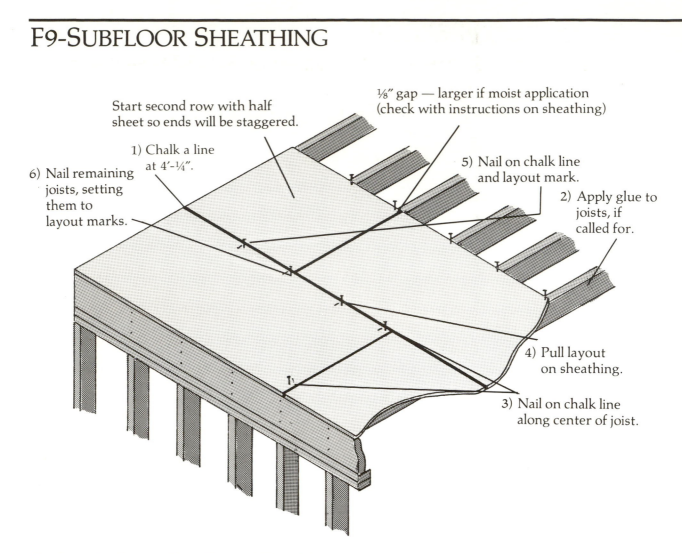

Start second row with half sheet so ends will be staggered.

1) Chalk a line at 4'-¼".

⅛" gap — larger if moist application (check with instructions on sheathing)

6) Nail remaining joists, setting them to layout marks.

5) Nail on chalk line and layout mark.

2) Apply glue to joists, if called for.

4) Pull layout on sheathing.

3) Nail on chalk line along center of joist.

Six Steps for Setting First Sheet of Sheathing

1) Chalk a line 4'-¼" in from one rim to its opposite, perpendicular to the joists.
2) Apply glue to joists, if called for. Be sure to nail sheathing before glue dries.
3) Center sheathing to last joist on the chalk line and nail.
4) Pull joist layout from corner of sheathing and mark sheathing.
5) Nail sheathing to joist next to rim joist along chalk line and layout mark.
6) Set remaining joists to layout marks and nail.

Setting Second Sheet

Set to chalk line and layout mark.

Setting Second Row and Remaining Sheets

Set to existing sheets allowing ⅛" gaps.
Stagger sheet ends on joists.
Make sure rim joists are straight before they are nailed.

Chapter Four

WALL FRAMING

There are many ways to frame walls, but it is always good to follow an organized sequence. This **sixteen-step sequence** has been developed over years of framing. Following these steps will help you and your crew work efficiently and eliminate errors. It will also ensure consistency from framer to framer. For example, if you have to leave a wall in the middle of framing it to go to another task, another framer can easily pick up where you left off and proceed without having to check every nail to see what you have done.

Keep in mind, walls must be *square*, *plumb*, and *level*. Measure accurately, cut straight, and nail tight.

W1-Spread Headers

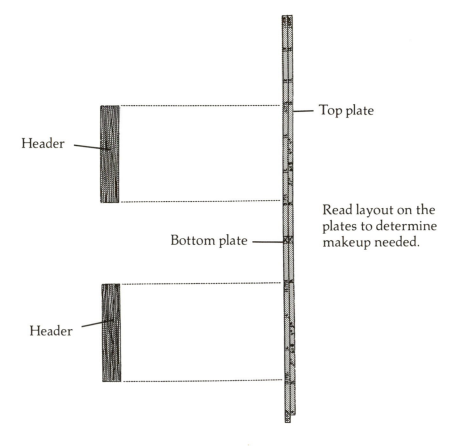

Header

Header

Top plate

Bottom plate

Read layout on the plates to determine makeup needed.

W2-Spread Makeup

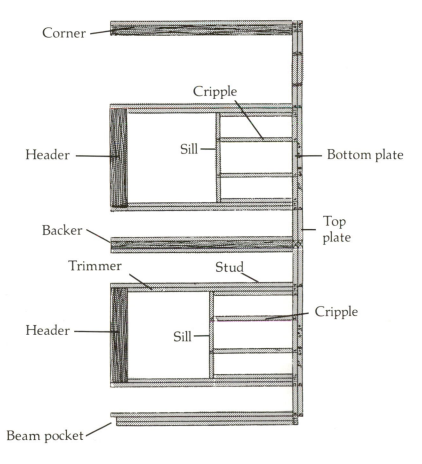

Corner

Cripple

Header

Sill

Bottom plate

Top plate

Backer

Trimmer

Stud

Header

Sill

Cripple

Beam pocket

W3-Spread Studs
W4-Nail Headers to Studs
W5-Nail Top Plate to Studs and Headers
W6-Nail Bottom Plate to Studs
W7-Nail Double Plate to Studs

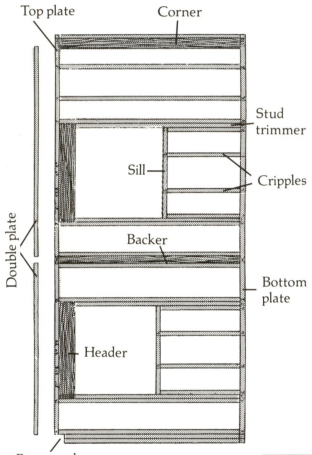

Top plate

Corner

Stud trimmer

Sill

Cripples

Double plate

Backer

Bottom plate

Header

Beam pocket

W8-Square Wall

Exterior Wall

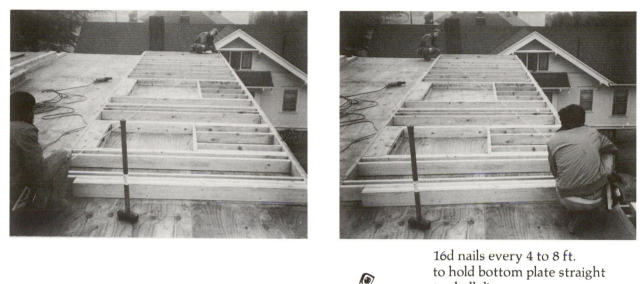

16d nails every 4 to 8 ft. to hold bottom plate straight to chalk line.

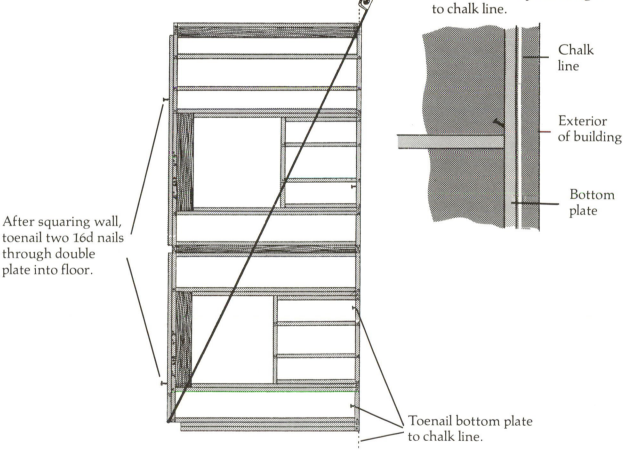

Chalk line

Exterior of building

Bottom plate

After squaring wall, toenail two 16d nails through double plate into floor.

Toenail bottom plate to chalk line.

To square a wall, secure the bottom plate as shown, then move the top of the wall until the diagonal dimensions are equal. Once the wall is square, secure it with two nails through the double plate into the floor.

Nail on the inside of the bottom plate so the nails will hold the wall in position while it is being stood.

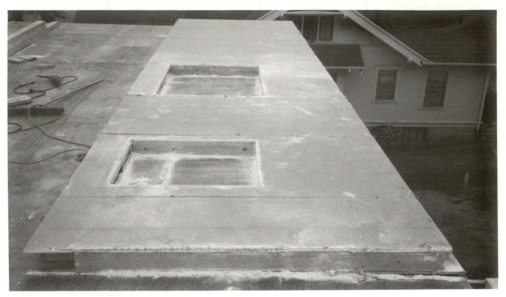

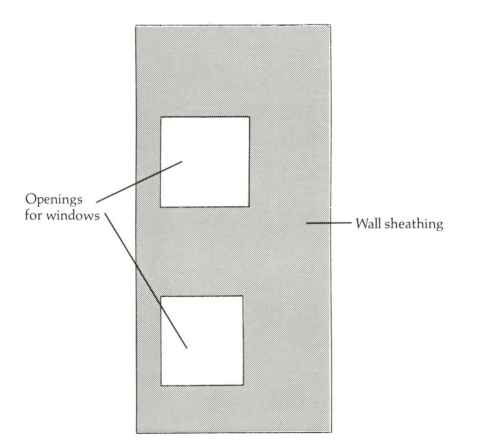

Openings for windows

Wall sheathing

Cover the entire wall with sheathing, then rout window and door openings with a panel pilot router bit. Save the leftover pieces of sheathing for small areas and filling in between floors.

If the first floor exterior walls can be reached from the ground, then the sheathing is not installed until after the walls are plumb and lined. This eliminates the potential problem of a square wall sitting on a foundation that is not level.

W10-INSTALLING NAIL-FLANGE WINDOWS IN WALL BEFORE WALL IS STOOD UP

a. Check plans for correct window.

b. Check window opening for protrusions (nails, wood splinters, etc.) that might hold window away from edge.

c. Set window in opening, making sure window is right side up.

d. Slide window to each end of opening and draw a line on the sheathing with a pencil along the edge of window.

e. Center windows in marks you have just drawn.

f. Nail window sides and bottoms, using appropriate nails.

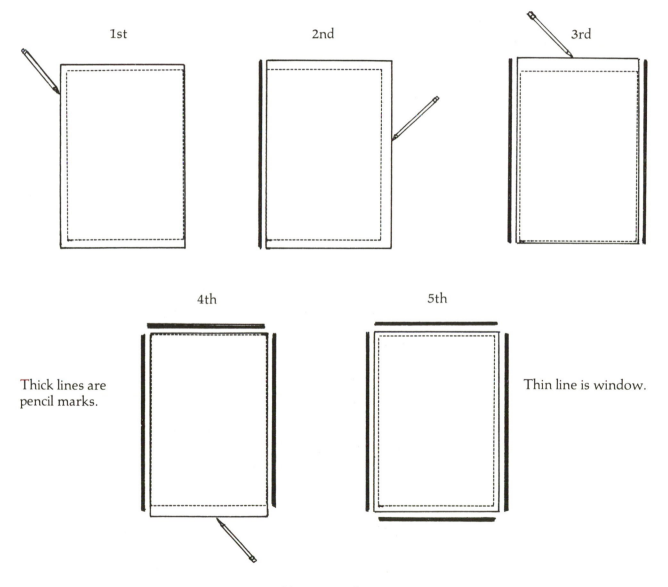

Thick lines are pencil marks.

Thin line is window.

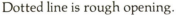

Dotted line is rough opening.

W11-STAND WALL
W12-SET BOTTOM PLATE
W13-SET DOUBLE PLATE
W14-SET REVEAL
W15-NAIL WALL

W14

Nail the end stud to set the reveal* in the middle of the wall where the two walls join. The reveal in the middle of the wall should be the same as the top and bottom reveal.

*The reveal is the amount of space on the corner stud of a wall after another wall is joined to it.

W15

Nail walls together and nail bottom plate.

W13

Set double plate.
Make sure top plates are down tight to studs.

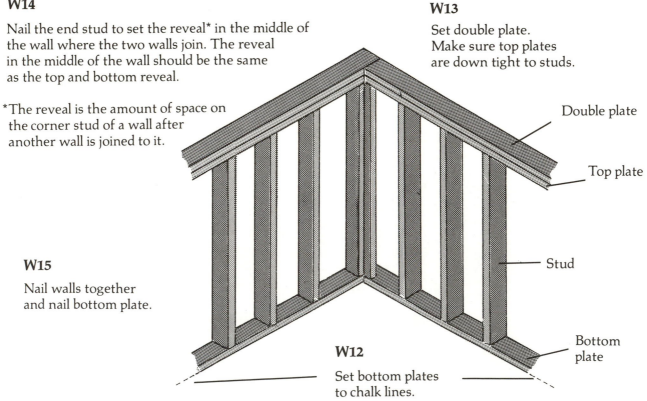

Double plate

Top plate

Stud

Bottom plate

W12

Set bottom plates to chalk lines.

W16-PLUMB AND LINE

"Plumb and line" is the process of making the walls straight and true.

"Plumbing" is the use of a level to set the ends of the walls plumb or perfectly upright.

"Lining" is using a tight string attached to the top of a wall to use as a guide for straightening it.

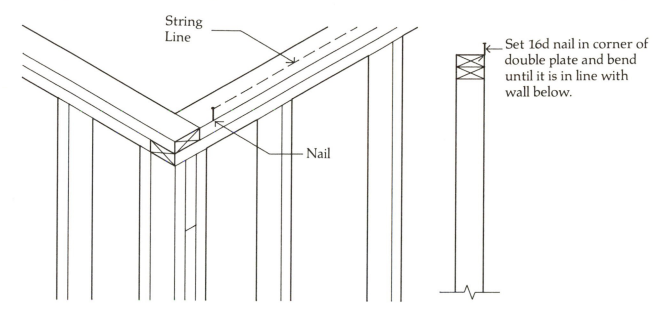

Set nails at either end of wall as shown, and then string line tightly between them, adjusting the line so that it is about ½" above the double plate. Wall should be moved in or out to align with string.

The walls are braced with 2 × 4 lumber to hold them and, if necessary, make them plumb and straight.

If a wall already is sheathed and in place, but not plumb, correct it if it is more than ¼" out of plumb for standard height walls.

Pulling Brace

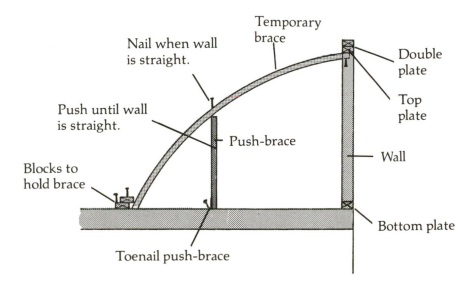

Racking Brace

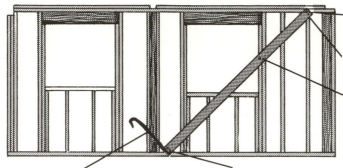

Do not let brace protrude above double plate.

Two 16d nails in top plate.

One 16d nail in center if it is a bearing wall, exterior wall, or will be used for walking up.

Use crowbar to rack brace and wall. Turn brace around if wall needs to be racked in other direction.

Two 16d nails into bottom plate.

Chapter Five

ROOF FRAMING

Framing a roof is the most difficult aspect of the trade, and the ability to construct a roof is a real test of your framing skills. Planning a roof requires knowledge of some rather complicated geometry, and building it means working in often awkward positions, at heights, with long and heavy pieces of lumber. The hardest part is calculating the lengths and angles for cutting rafters. Don't expect to be able to read this chapter and go out and frame a roof. Being able to frame a roof takes time, experience, and many wrong cuts. Don't get discouraged by mistakes and setbacks on the first few roofs you do. Knowledge comes from struggling through failure: learn from those mistakes.

This manual presents three methods for finding the rafter length. The first is the "Pythagorean Theorem." To use this method you need to have a basic knowledge of geometry and access to a calculator with a square root key. A framing square with a rafter table imprinted on it is required for both the "Framing Square Rafter Table" method and the "Framing Square Stepping" method.

Prefabricated trusses, factory-made from the architect's specifications, require, of course, no calculations on your part; and being uniform in size, are somewhat easier to work with than ridgeboards and rafters. Still, they are heavy and you are working at heights. You must be constantly aware of what you and the other two crew members are doing. Planning and teamwork are required. Before you have the trusses set on the roof, make sure you spend some time thinking about the easiest way for them to go up. If they get set on the roof in a disorganized manner, it can mean much extra work and a lot of aggravation moving them around on top of the walls.

ROOF FRAMING TERMS

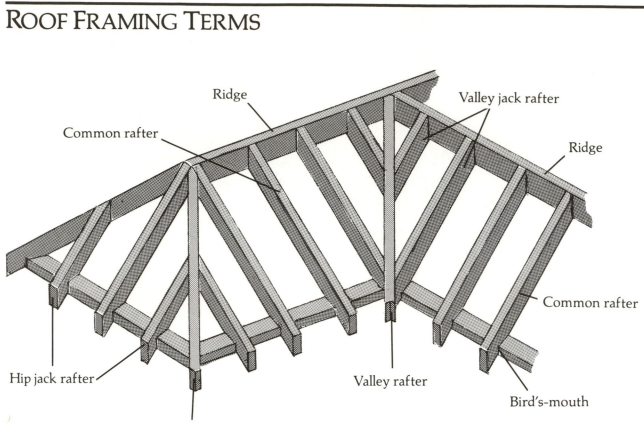

Ridge

Valley jack rafter

Common rafter

Ridge

Common rafter

Hip jack rafter

Valley rafter

Bird's-mouth

Hip rafter

ROOF STYLES

Gambrel

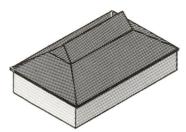

Dutch hip

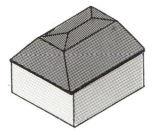

Mansard

Gable

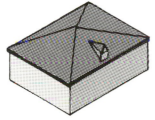

Pyramid with dormer

Flat

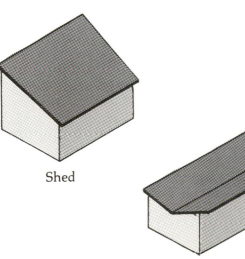

Shed

Butterfly

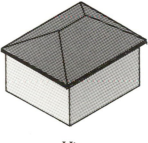

Hip

R1-FINDING THE LENGTHS OF COMMON RAFTERS

There are three methods for finding common rafter lengths. Study them all and use the one that works best for you.

A. Phythagorean Theorem

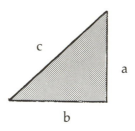

B. Framing-Square Rafter Table

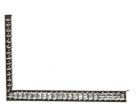

C. Framing-Square Stepping Method

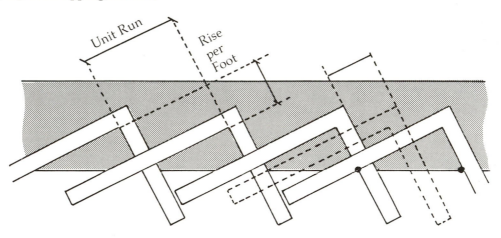

A-Pythagorean Theorem

Pythagoras was an ancient Greek philosopher and mathematician. His famous theorem states that the square of the hypotenuse of a right triangle is equal to the sum of the squares of the two other sides. Thus:

$$A^2 + B^2 = C^2$$

In roof framing, A = the Rise, B = the Run, and C (the hypotenuse) = the Rafter Length.

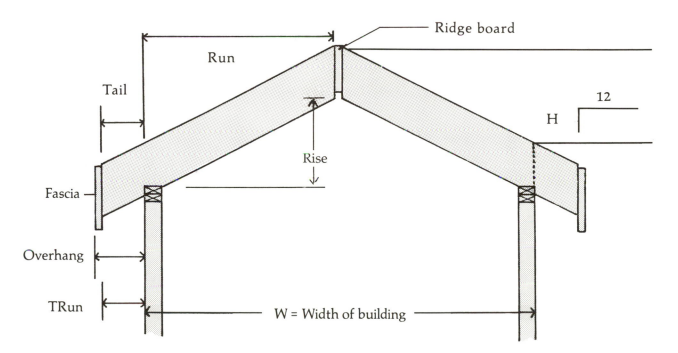

Run = ½ building width – ½ ridge board width
H = is given on plans = the amount of rise per foot of run

Rafter Cut Length = Rafter Length + Rafter Tail Length

Finding Rafter Length

First, find the run by using the following formula:

    Run = ½ building width – ½ ridge board width

Second, find the rise by using the following formula:

    Rise = H / 12 × Run

Third, find the Rafter Length by using the following formula

    Rafter Length = $\sqrt{(Rise \times Rise) + (Run \times Run)}$

A-Pythagorean Theorem *(continued)*

To apply this formula, multiply Rise × Rise, and then Run × Run. Add the two sums, then press the square root key on your calculator. The result is the Rafter Length.

Finding Rafter Tail (T) Length

First, find the TRun by using the following formula:

TRun = Overhang – Fascia

Second, find the TRise using the following formula:

TRise = H/12 × TRun

Third, find the Rafter Tail Length by using the following formula:

Rafter Tail Length = $\sqrt{(\text{TRise} \times \text{TRise}) + (\text{TRun} \times \text{TRun})}$

Note: Be sure to mark crowns on rafters prior to measuring and cutting. Crowns are always up.

Example: Finding Rafter Cut Length

Rafter Length:

Let the pitch be $4\overline{\smash{\big)}12}$ and the building width be 20′ and the ridge board be 1½″ thick.

Step 1: Run = ½ (20′) – ½ (1½″) = 9′-11¼″

Step 2: Rise = 4/12 × 9′-11¼″
        = .3333 × 119.25″
        = 39.75
        = 39¾″

Step 3: Rafter Length = $\sqrt{(119.25 \times 119.25) + (39.75 \times 39.75)}$
        = $\sqrt{14{,}220.56 + 1{,}580.06}$
        = $\sqrt{15{,}800.62}$
        = 125.70
        = 125¹¹/₁₆″

Rafter Tail Length:

Let overhang be 2′ and fascia be 1½″.

Step 1: TRun = 2′ – 1½″ = 1′-10½″
Step 2: TRise = 4/12 × 1′ – 10½″
        = .3333 × 22.5″ = 7.49
        = 7.49
        = 7½″

Step 3: Rafter Tail Length = $\sqrt{(7.49 \times 7.49) + (22.5 \times 22.5)}$
        = $\sqrt{56.10 + 506.25}$
        = $\sqrt{562.35}$
        = 23.71
        = 23¹¹/₁₆″

Rafter Cut Length = 125¹¹/₁₆ + 23¹¹/₁₆
        = 149⅜″

B-Framing-Square Rafter Table

Steps

1) Find pitch of roof from plans.
2) Find pitch on framing-square inch line.
3) Find multiplication factor below pitch number. ────
4) Find run of rafter and multiply by multiplication factor.
5) Subtract ½ thickness of ridge board to determine the rafter length.

Example

Let the pitch be $4\overline{\smash{)}^{12}}$ and the run be 10'.

Step 1 = 4
Step 2 = 4
Step 3 = 12.65
Step 4 = 10 × 12.65 = 126.5" ──────
Step 5 = 126.5" – ¾" = 125.75" rafter length

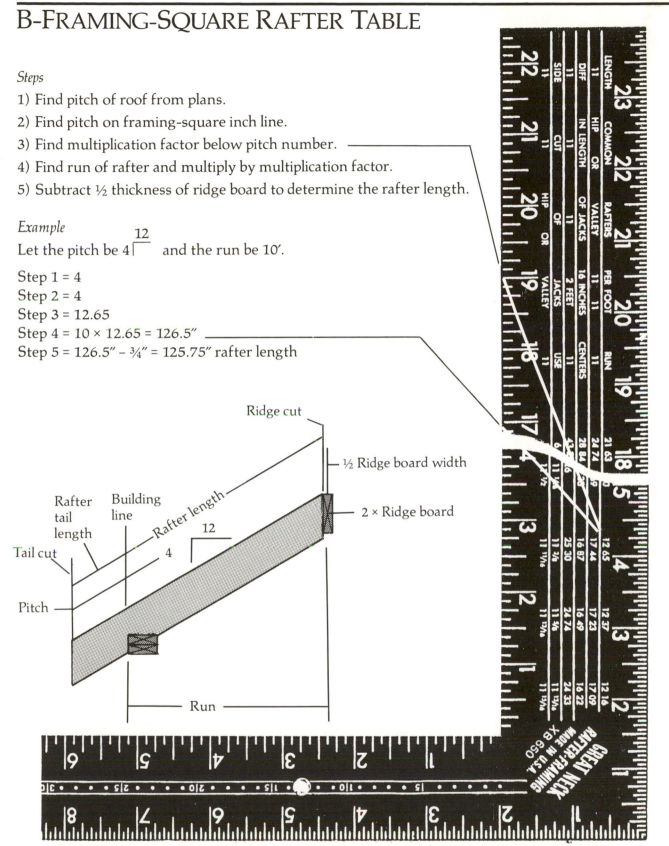

Rafter Tail Length for overhang is found using the same method as for the rafter length except that Step 5 (subtraction for ridge board) is eliminated. The rafter tail length is then added on to the rafter length to produce the rafter tail cut length.

C-FRAMING-SQUARE STEPPING METHOD

The pitch of the roof is given on the plans in this way: $6 \lceil \frac{12}{}$

The number 12 is constant and indicates 12 inches of run, or horizontal distance. The other number represents the rise and varies, depending on how steep the roof is. The example below indicates that for every 12″ of run, or horizontal distance, there is 6″ of rise, or vertical distance. The greater this number, the steeper the roof.

To find the rafter length, first lay the framing square on the rafter at 12″ on the blade and the amount of rise on the tongue. Once the framing square is set with stair nuts, just step off the amount of run along the rafter.

Example

Let the run equal 10′-4″

Let the overhang equal 1′-6″

1) Set framing square with stair nuts: the run, 12″, on the blade; the rise, 6′, on the tongue.

2) Step off 11′.

3) Perpendicular to the end lines mark 4″ (top), and 6″ (bottom). Place the square at those marks to draw the plumb lines for the ridge and tail cuts.

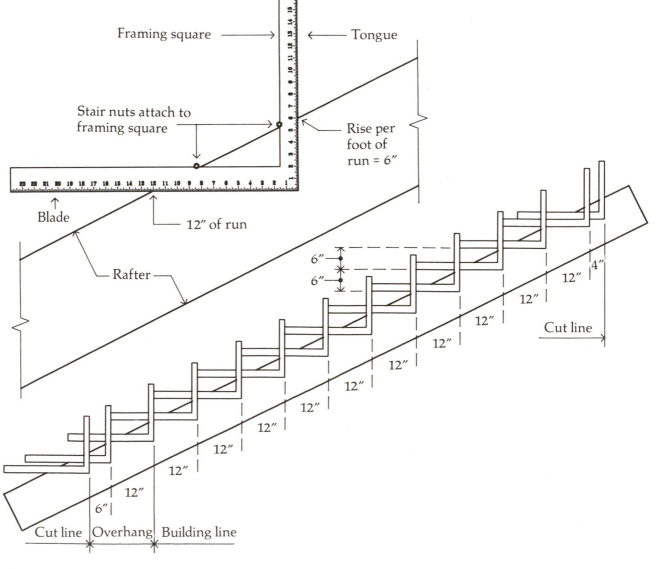

52

R2-Cut Common Rafter

A) Rafter cut length

B) Bird's-mouth

C) Angle cuts

A) Rafter Cut Length

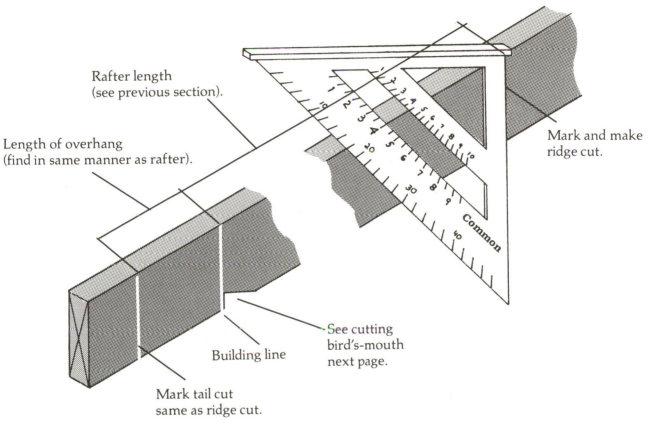

Rafter length (see previous section).

Length of overhang (find in same manner as rafter).

Mark and make ridge cut.

See cutting bird's-mouth next page.

Building line

Mark tail cut same as ridge cut.

Set speed square to H (which is given on plans).

H = the amount of rise per foot of run.

This speed square, as an example, is set to H = 4.

Shown on plans like this: 4 $\overline{12}$

Cut a pattern first and try it for fit before cutting all the rafters. A framing square can also be used to mark cut lines (see illustration in "Framing-Square Stepping Method.")

R2-Cut Common Rafter *(continued)*

B. Bird's-Mouth

Steps:

1) Mark rafter length.
2) Mark building line at rafter length for the correct pitch.
3) Mark parallel plumb line a distance equal to width of wall and toward interior of building.
4) Mark seat cut square (90°) from building line at rafter length and bottom of parallel line.
5) Cut bird's-mouth.

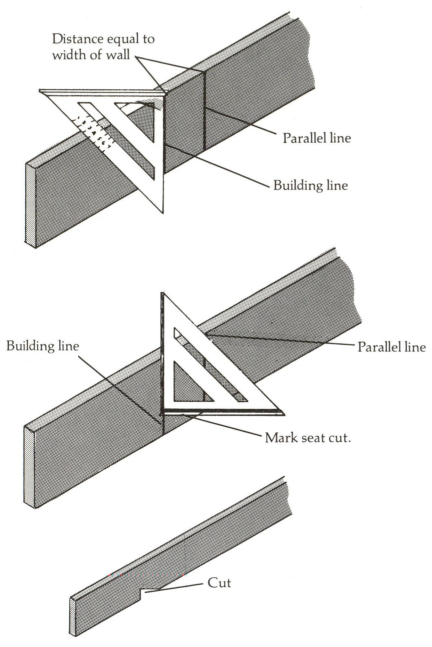

Distance equal to width of wall

Parallel line

Building line

Building line

Parallel line

Mark seat cut.

Cut

A framing square can also be used to mark a bird's-mouth. (See illustration in "Framing-Square Stepping Method")

R2-CUT COMMON RAFTER (continued)

C. Angle Cuts

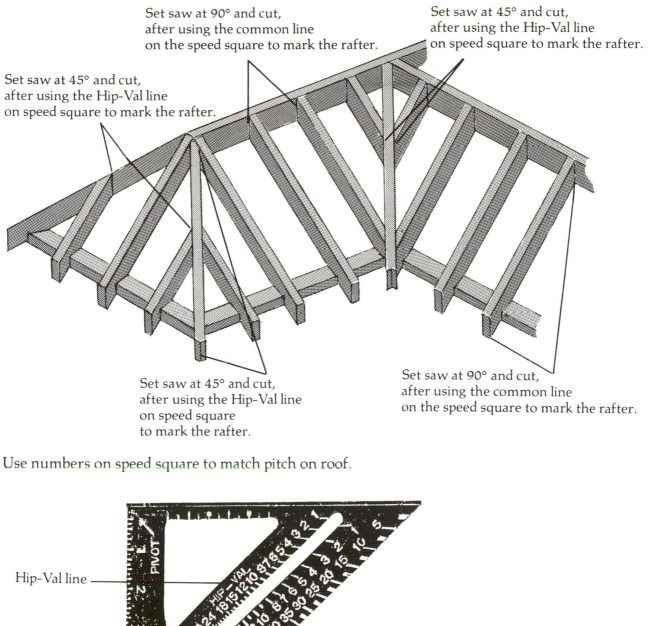

Set saw at 90° and cut, after using the common line on the speed square to mark the rafter.

Set saw at 45° and cut, after using the Hip-Val line on speed square to mark the rafter.

Set saw at 45° and cut, after using the Hip-Val line on speed square to mark the rafter.

Set saw at 45° and cut, after using the Hip-Val line on speed square to mark the rafter.

Set saw at 90° and cut, after using the common line on the speed square to mark the rafter.

Use numbers on speed square to match pitch on roof.

Hip-Val line

Common line

R3-SET RIDGE BOARD

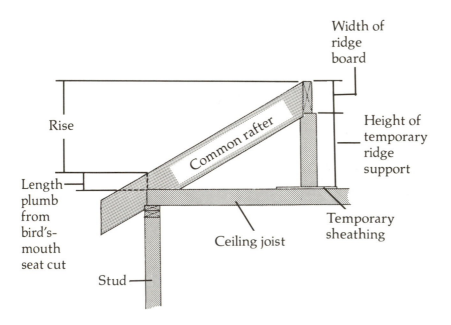

Ridge board

Temporary ridge support

Double plate

Top plate

Temporary sheathing

Stud

Ceiling joist

Height of Temporary Ridge Board Support

Width of ridge board

Rise

Common rafter

Height of temporary ridge support

Length plumb from bird's-mouth seat cut

Temporary sheathing

Ceiling joist

Stud

Height of temporary ridge support =
    Rise + length of plumb from bird's-mouth seat cut
        – width of ceiling joist
        – thickness of temporary sheathing
        – ridge board width

R4-Set Common Rafters

Nail three 16d nails through ridge board into rafter. Toenail when necessary.

Common rafter

Common rafter

Ridge board

Temporary ridge support

Temporary sheathing

Top plate

Ceiling joist

Studs

Double plate

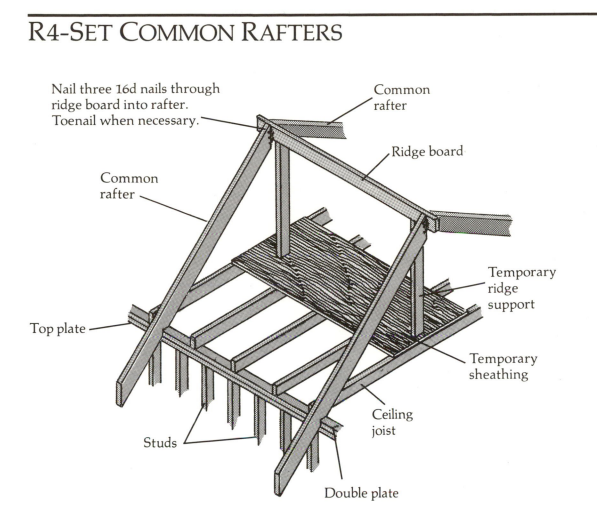

To start the ridge board layout, plumb a line up from a string held tight between roof layout marks on opposite walls. From that point, mark the ridge board layout to match the roof layout on the double plate.

Begin by setting the end rafters, as shown. Set the remaining rafters in the order that works best for you. As in lining a wall, attach a string in a similar manner along the edge of the ridge board. This will guide you in keeping a straight ridge board while you set the remaining rafters.

R5-LENGTH OF HIP AND VALLEY RAFTERS

The length of the hip and valley rafters can be found by using any of the three common rafter methods previously described, and then making adjustments for the run and the top and bottom cuts.

Adjustment for the Run

For every 12" of common rafter run, there is 16.97" (17" approx.) of run for hip and valley rafters.

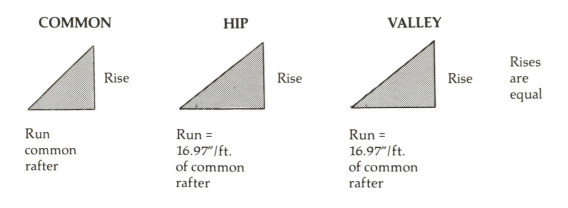

COMMON	HIP	VALLEY	
Rise	Rise	Rise	Rises are equal
Run common rafter	Run = 16.97"/ft. of common rafter	Run = 16.97"/ft. of common rafter	

Multiply the run in feet of the common rafter by 16.97" (17 is commonly used) to get the run of the hip or valley rafter.

Adjustments for the Top and Bottom Cuts

The cut mark will be made similar to the common rafter cut mark, except that the hip-val scale on the speed square will be used instead of the common scale to mark the line to cut. (See "Rafter Cut Length" and "Angle Cuts.")

If a framing square is used, use the same procedure shown previously, except use 17 instead of 12 along the blade of the framing square.

These procedures assume a hip or valley corner of 90 degrees.

R6-Cut Hip Rafter

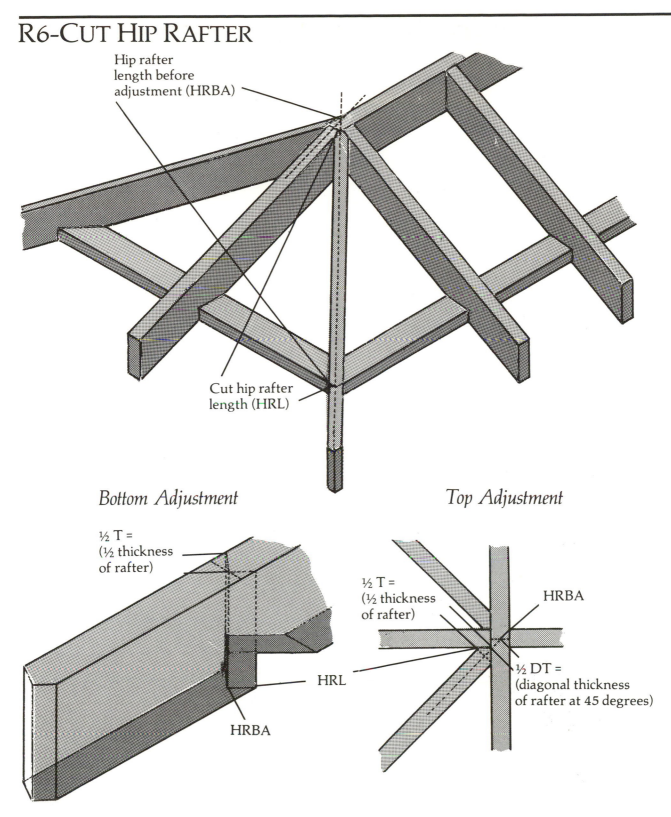

Hip rafter length before adjustment (HRBA)

Cut hip rafter length (HRL)

Bottom Adjustment

½ T = (½ thickness of rafter)

HRL

HRBA

Top Adjustment

½ T = (½ thickness of rafter)

HRBA

½ DT = (diagonal thickness of rafter at 45 degrees)

Hip Rafter Length (HRL) = Hip rafter length before adjustments (HRBA) – ½ diagonal thickness (DT) of rafter – thickness (T) of rafter (½ thickness top and bottom combined).

or

HRL = HRBA – ½ DT – T

These figures are based on a 90-degree building corner.

R6-CUT VALLEY RAFTER

Valley rafter length before adjustment (VRBA) (see "Length of")

Valley rafter length (VRL)

Bottom Adjustment   *Top Adjustment*

½ T = (½ thickness of rafter)

½ DT = (diagonal thickness of rafter at 45 degrees)

½ T = (½ thickness of rafter)

Valley Rafter Length = Valley rafter length before adjustment – ½ diagonal thickness of rafter. (The ½ thickness factors cancel each other out.)

or

VRL = VRBA – ½ DT

These figures are based on a 90-degree building corner.

R7-Set Hip and Valley Rafters

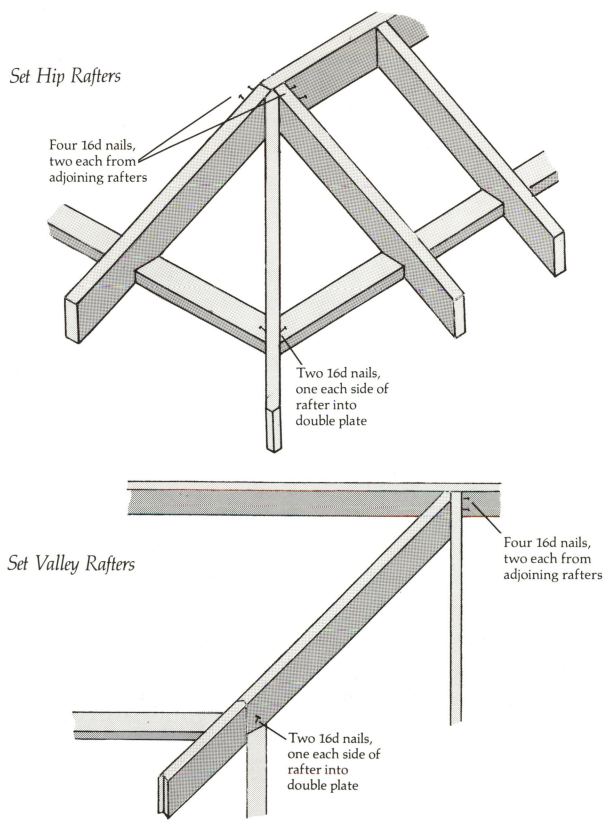

Set Hip Rafters

Four 16d nails, two each from adjoining rafters

Two 16d nails, one each side of rafter into double plate

Set Valley Rafters

Four 16d nails, two each from adjoining rafters

Two 16d nails, one each side of rafter into double plate

Note: If hip or valley rafters do not fit, first check your length calculations, then check walls for plumb and the ridge board for correct height and for plumb.

R8-Set Jack Rafters

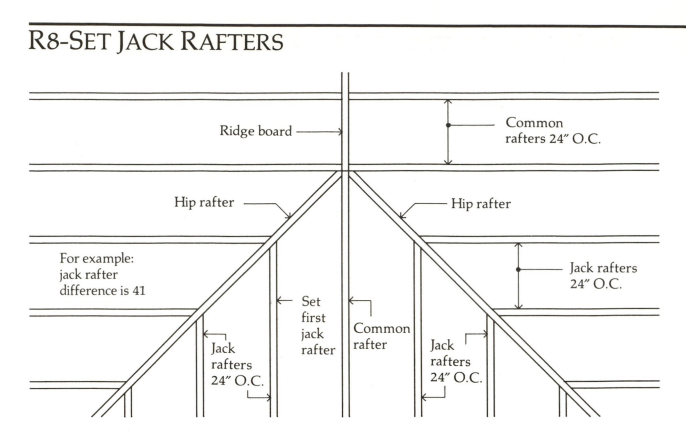

Set first jack rafter on 16" or 24" spacing with common rafters.

Measure length from common rafter to first jack rafter and then use standard jack rafter differences, as given in framing square table, to measure lengths of remaining jack rafters along the hip rafter.

Framing Square Table

					18	17	16
LENGTH	COMMON	RAFTERS	PER FOOT	RUN	21 63	20 81	20
"	HIP OR	VALLEY	" "	"	24 74	24 02	23 32
DIFF	IN LENGTH	OF JACKS	16 INCHES	CENTERS	28 7/8	27 3/4	26 11/16
"	"	"	2 FEET	"	43 1/4	41 5/8	40
SIDE	CUT	OF	JACKS	USE	6 11/16	6 15/16	7 3/16
"	"	HIP OR	VALLEY	"	8 1/4	8 1/2	8 3/4

For example: jack rafter difference is 41

R9-BLOCK RAFTERS AND LOOKOUTS
R10-SET FASCIA

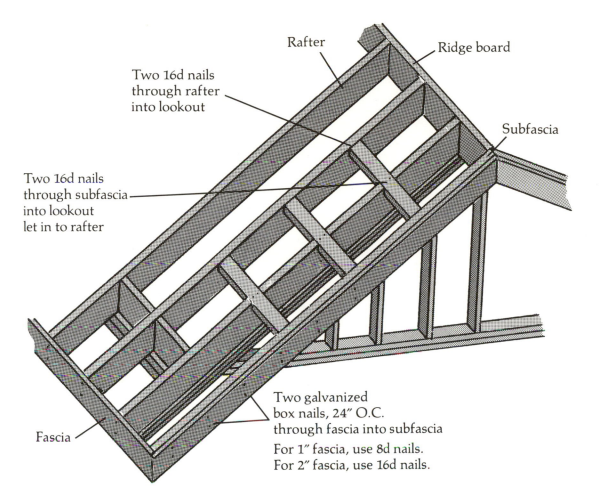

Rafter

Ridge board

Two 16d nails through rafter into lookout

Subfascia

Two 16d nails through subfascia into lookout let in to rafter

Two galvanized box nails, 24" O.C. through fascia into subfascia

For 1" fascia, use 8d nails.
For 2" fascia, use 16d nails.

Fascia

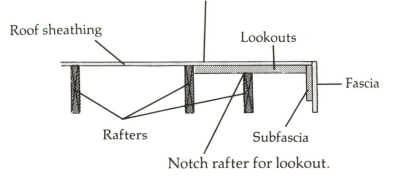

If overhang is exposed underneath, use ccx plywood for exposed portion.

Roof sheathing

Lookouts

Fascia

Rafters

Subfascia

Notch rafter for lookout.

R11-SHEATHING

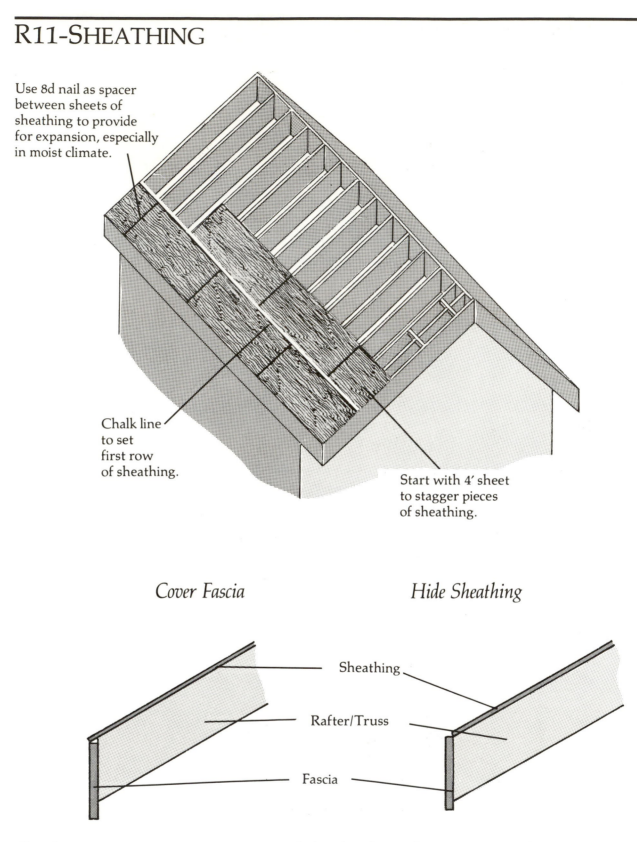

Use 8d nail as spacer between sheets of sheathing to provide for expansion, especially in moist climate.

Chalk line to set first row of sheathing.

Start with 4′ sheet to stagger pieces of sheathing.

Cover Fascia

Hide Sheathing

Sheathing

Rafter/Truss

Fascia

Sheathing covers fascia.

Fascia hides sheathing edge. Fascia is attached so that its top outer edge is in line with the plane of the top surface of the roof sheathing.

T1-Spread Trusses

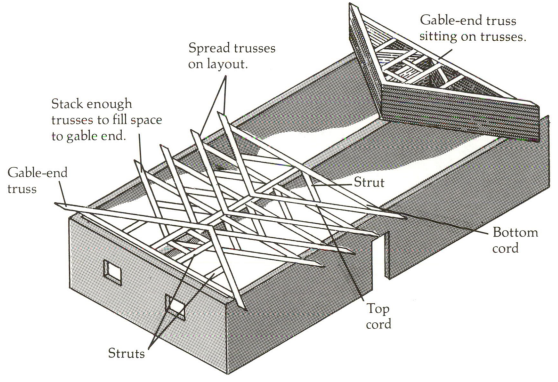

Gable-end truss sitting on trusses.

Spread trusses on layout.

Stack enough trusses to fill space to gable end.

Gable-end truss

Strut

Bottom cord

Top cord

Struts

When trusses are delivered in stacks, they should be set on the roof to allow for easy spreading. The gable ends should be on top because they go up first. The direction of the ridge is important so they can be spread and tilted up easily.

T2-SHEATH GABLE ENDS

When sheathing gable-end truss, check plans to see if a vent opening is shown.

- Set gable end in place.
- Center gable end on wall.
- Toenail through bottom of truss into plate so gable end is on chalk line 1½" in from outer edge of double plate.

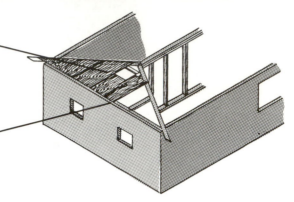

T3-SET GABLE ENDS

Toenail 16d nails 24" O.C. to hold gable end while setting gable-end truss.

1½" chalk line

Truss

Wall

Temporary braces

Temporary braces

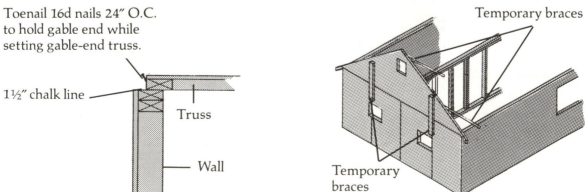

T4-Roll Trusses

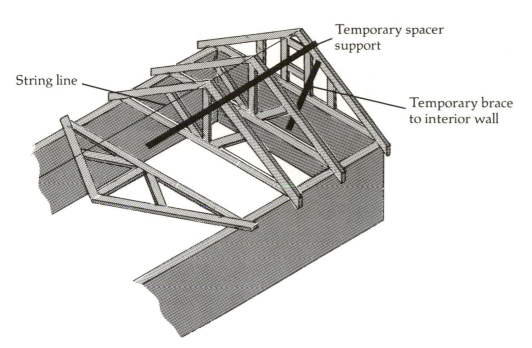

String line

Temporary spacer support

Temporary brace to interior wall

STEPS

1) String line from center of gable ends.

2) Lift single truss into place.

3) Center truss on string line.

4) Nail truss to exterior wall on layout.

5) Nail through temporary spacer support near ridge of truss and on layout marked on temporary spacer support.

6) Set six trusses. Then check gable end for plumb and put permanent brace on gable end. Permanent brace should connect top of gable end to an interior wall or a cross support running between the trusses.

7) Every eight trusses, put an additional brace on trusses. Refer to truss specifications for additional braces.

The block trusses, lookouts, fascia, and sheathing installations described in the previous pages are shown below, in actual project photos.

T5-Block trusses   **T5-Block trusses**   **T7-Fascia**

T6-Lookouts
T8-Sheathing

Chapter Six

DOORS, WINDOWS, AND STAIRS

Doors and windows are two of the few finish items that framers handle. It is important that time and care are taken so that they are installed in a proper, professional manner. Put your framing hammer in the tool box and use, instead, a lighter, smoother-faced trim hammer and a nail set.

Exterior, prehung doors are the type covered in this chapter. They are the ones framers most commonly work with, and most of the skills involved in hanging them will carry over to the hanging of interior doors. The first door you hang on any job will give you the most difficulty. If you have more than one door to hang, do them one after the other; each door will go in a little easier than the one before.

Nail-flange windows and sliding glass doors will vary depending on the manufacturer. You will find here the basic principles of their installation. Use common sense and follow the directions provided, and you should have little trouble installing these units.

Stairs represent one of the more difficult challenges to a framer's skill. As in roof framing, the geometry is a bit complicated, but taken step-by-step, the logic soon becomes clear, leading to successful execution of the plans. There are many different stair designs. The stair layout described in this chapter is typical. Be aware that the dimensions given on the plans do not always allow for enough headroom. Always check headroom and other dimensions by taking accurate on-site measurements.

A calculator is handy, some might say necessary, for finding the rise and tread dimensions when not given on the plans. Always double- or triple-check your calculations. Remember: measure twice, cut once. The finish floors at the top and bottom of stairs are often different. When cutting stair stringers, don't forget to check the plans for such differences and then check the height of your top and bottom risers to allow for them.

DOOR FRAMING TERMS

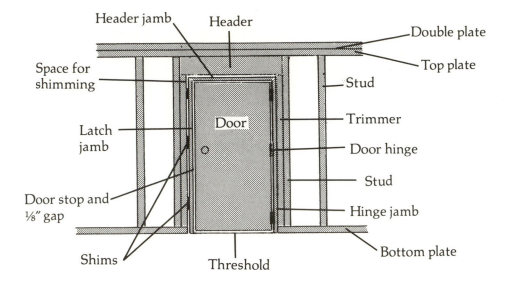

STEPS FOR INSTALLATION OF EXTERIOR PREHUNG DOORS

Steps

1) Read instructions that come with door.

2) Check plans for direction of door swing.

3) Check threshold for level (shim under hinge jamb if necessary).

4) Nail hinge jamb (do not set nails), and plumb both directions. Shim behind jamb if necessary to obtain plumb or if door needs to be centered.

5) Nail latch jamb using shims. Set so there is ⅛" margin between door and jamb and so top and bottom of door hit door stop at the same time when closing. Lock set holes also must line up.

6) Check door for total fit.

7) Nail door with six 16d galvanized casing nails in hinge jamb, two at each hinge, and four in latch jamb (one each at the top and bottom and two in the middle). This includes nails used in steps 4 and 5.

8) Set nails and break or cut shims.

9) Check again for total fit.

STEPS FOR INSTALLATION OF EXTERIOR PREHUNG DOORS *(continued)*

1. READ INSTRUCTIONS

Usually there are instructions that come with the door. Check the instructions over for anything you might need to know.

2. CHECK PLANS

Check the building plans to find the direction of the door swing.

3. CHECK THRESHOLD

Check the threshold for level. Shim under hinge jamb if necessary.

One 16d galvanized casing nail at each hinge.
DO NOT SET NAILS

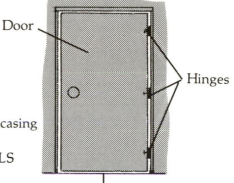

Door

Hinges

Threshold

Shim under hinge jamb if floor is low.

4. NAIL HINGE JAMB

Nail hinge jamb tight to trimmer with one 16d galvanized casing nail at each hinge. Do not set nails. Plumb both directions. Shim behind jamb if necessary to obtain plumb or if door needs to be centered in opening.

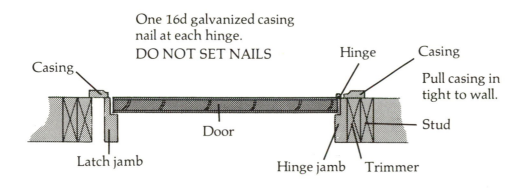

One 16d galvanized casing nail at each hinge.
DO NOT SET NAILS

Casing

Hinge   Casing

Pull casing in tight to wall.

Door

Stud

Latch jamb

Hinge jamb   Trimmer

STEPS FOR INSTALLATION OF EXTERIOR PREHUNG DOORS *(continued)*

5. NAIL LATCH JAMB

Before nailing the latch jamb, make sure of the following three items:
1) Continuous gap between door and jamb is ⅛".
2) Door touches latch jamb equally at top and bottom.
3) The lockset hole in door and in the latch jamb line up.

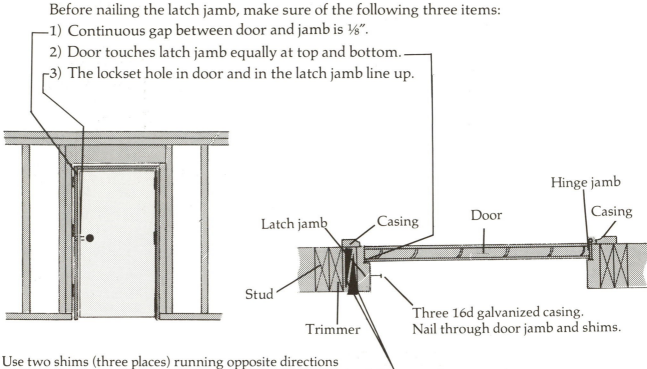

Use two shims (three places) running opposite directions so door jamb remains square. Shingles are commonly used for shims.

6. CHECK DOOR

Check door for total fit:
1) Gap around door is even.
2) Door and latch jamb align.
3) Door closes smoothly, no binding.
4) Lockset holes line up.
5) Door closes tight, no rattle.

7. NAIL DOOR

Six – 16d galvanized casing nails in hinge jamb — two at each hinge.

Four – 16d galvanized casing nails in latch jamb — one each top and bottom and two near latch. This includes nails used in steps 4 and 5.

8. SET NAILS AND TRIM SHIMS

9. FINAL CHECK

Check again for total fit (see step 6).

INSTALLATION OF NAIL-FLANGE WINDOW AFTER WALL IS UP

1) Set in place (from inside if possible).
2) Place temporary shims under bottom of window. Equalize space at top and bottom of window. Shims are usually ⅛" to ¼".
3) Make gap the same between window frame and trimmer on each side.
4) Nail top corners from outside.
5) Nail one bottom corner (do not set all the way).
6) Place window slider in and check to see if the gap between the window slider and window frame is the same top to bottom.
7) Check. If the gap is not equal, check both the rough opening and the window for square and adjust accordingly.
8) Finish nailing. Make sure gap is equal top, bottom, and middle. (Do not nail top of the window.)

View From Inside

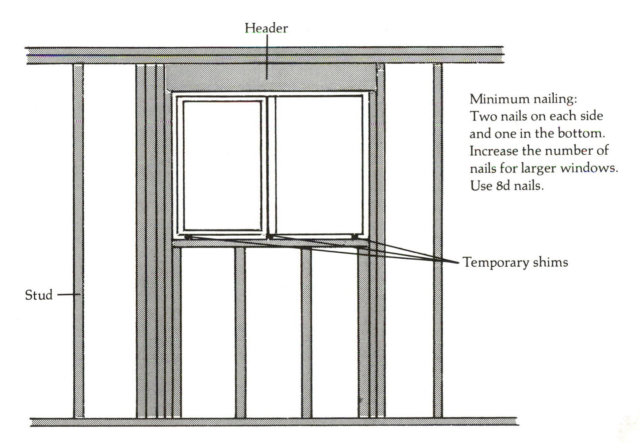

Header

Minimum nailing:
Two nails on each side
and one in the bottom.
Increase the number of
nails for larger windows.
Use 8d nails.

Temporary shims

Stud

STEPS FOR INSTALLATION OF SLIDING GLASS DOORS

1) **INSTRUCTIONS**
   Read and follow carefully the instructions that come with the door. Never assume what you do not know.

2) **SEAL THRESHOLD**
   Use neoprene or similar sealing compound to seal the threshold.

3) **PLACE DOOR**
   Place the door in position.

4) **CENTER TOP**
   Center the top of the door in rough opening.

5) **NAIL TOP CORNERS**
   Nail each corner of the top of the door through nail flange (do not set nails).

6) **ADJUST DOOR**
   Adjust the door frame so that the space between the door frame and wall trimmers is equal. Check for plumb with a level and adjust if necessary.

7) **NAIL COMPLETE**
   Close the door and latch it. Then nail off the sides using four 8d nails on each side. Do not nail top of door.

8) **ADJUST DOOR**
   Adjust the slider part of the door if necessary. Usually there is an adjustment screw at the bottom of the door. Tighten this screw to close a gap between the door and the jamb at the top of the door, or loosen to increase the gap.

9) **SCREWS**
   If screws come with the door, shim and tighten screws in the sides and bottom. Use predrilled holes.

INSTALLATION OF SLIDING GLASS DOORS

Sliding Glass Door — from Inside

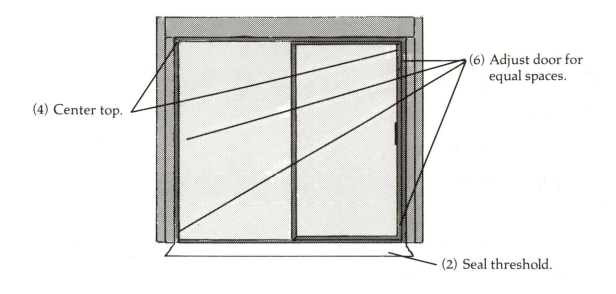

(4) Center top.

(6) Adjust door for equal spaces.

(2) Seal threshold.

Sliding Glass Door — from Outside

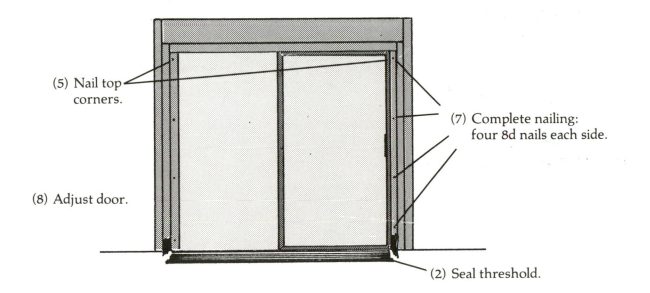

(5) Nail top corners.

(7) Complete nailing: four 8d nails each side.

(8) Adjust door.

(2) Seal threshold.

STAIR INSTALLATION

The three main dimensions in stair building are those for risers, treads, and headroom. The riser height and the tread width are usually given on the plans. You can generally use the tread width given on the plans. The riser height, however, is often not accurate enough to use.

Stair Installation Steps

1) **MEASURE HEIGHT**
Measure the height of the stairwell from finish floor to finish floor.

2) **FIND RISER HEIGHT**
Divide the height of the stairwell by the number of risers shown on the plan to determine the riser height. Be careful to consider the finish floor heights, which may differ top and bottom.

3) **FIND TREAD WIDTH**
Check plans for tread width.

4) **CHECK HEADROOM**
Chalk a line from edge of nosing at top of stairs to edge of nosing at bottom of stairs (see "Checking Stair Headroom" in the next pages of the section). Check for minimum clearance of 6'8" to finish straight up from line to bottom of headroom.

5) **DRAW AND CUT STRINGERS**
(See "Checking Stair Headroom.")

6) **CUT**
Cut stringer spacers, treads, and risers.

7) **NAIL STRINGER SPACER**
Nail stringer spacer to stringer. Spacer leaves clearance for applying wall finish.

8) **SET STRINGERS**
   a) At top deck measure down riser height plus tread thickness and mark for top of stringer.
   b) Set stringers to mark.
   c) Check stringers for level by placing a tread on top and bottom and checking level side to side and front to back.
   d) Adjust stringers for level.
   e) Nail stringers.

9) **NAIL RISERS**

10) **GLUE AND NAIL TREADS**

FINDING RISER AND TREAD DIMENSION WHEN NOT GIVEN ON PLANS

If the riser and tread dimensions are not given on the plan, then you need to calculate them. To do this you should consider the following points:

- You want the steps to feel comfortable.
- When walking up steps, a person's mind determines the height of the riser based on the first step. Make sure all risers and treads are equal, so the stairs will not cause people to fall.
- The lower the riser, the longer the tread needs to be to feel comfortable.
- Common dimensions for riser and tread are 7" rise and 10½" tread.
- Use the following three rules to check to see if your stair dimensions are in the comfortable range.
  — Rule 1: Two risers and one tread added should equal 24" to 25".
  — Rule 2: One riser and one tread added should equal 17" to 18".
  — Rule 3: Multiply one riser by one tread and the result should equal 71" to 75".

IMPORTANT STAIR CODE REGULATIONS 1988 UBC CODES SECTION 3306 STAIRS

Width – 36" minimum – with occupant load of 49 or less.
        44" minimum – with occupant load of 50 or more.

Rise   – 4" minimum
        7" maximum – 8" for occupant load of less than 10.

Tread – 11" minimum – 9" for occupant load of less than 10.

Riser height and tread length variance – ⅜" maximum.

Headroom – 6'8" minimum, measured vertically from a plane parallel and tangent to the stairway tread nosing to the soffit above.

CHECKING STAIR HEADROOM

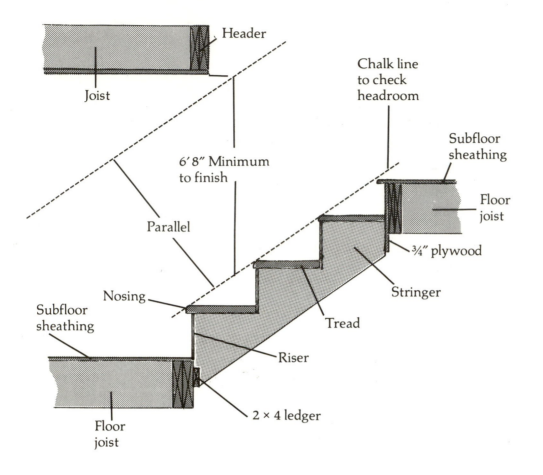

Checking Headroom Before Stairs Are Built

First measure up from the subfloor sheathing at the bottom of the stair the height of one riser; then measure back the distance of the nosing and make a mark. From that mark, measure out parallel to the subfloor a distance equal to the combined width of the total number of treads and measure perpendicular to the subfloor the height of the total number of risers and make a second mark. Chalk a line between these two marks. From this line you can measure up to see if there are any obstructions in the headroom.

MARKING STAIR STRINGERS

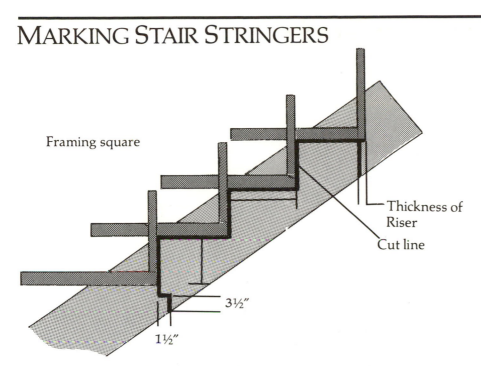

Framing square

Thickness of
Riser

Cut line

3½"

1½"

Measure treads and risers using framing square.

Subtract thickness of riser from top.

Notch bottom for ledger or stop plate. Notches differ (see above and the following page).

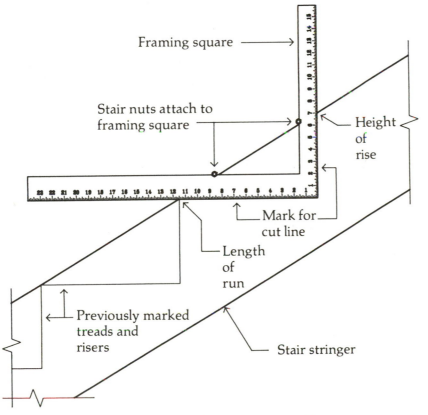

Framing square →

Stair nuts attach to
framing square

Height
of
rise

Mark for
cut line

Length
of
run

Previously marked
treads and
risers

Stair stringer

Care must be taken when marking the top and bottom steps. The thickness of the stair tread and the type of finish flooring on both the tread and the floor must be considered so that all the risers will be the same.

For the top tread be sure to figure in the riser so that the treads and nosings are all equal.

STANDARD STAIR (To be Carpeted)

Adapt for use when plans do not give details.

2 × 12 Stair tread
- Router nosing with ½" round router bit.
- Glue and nail three 16d nails each stringer for single residence use.
- Glue and nail four 16d nails each stringer for multi-residence use.

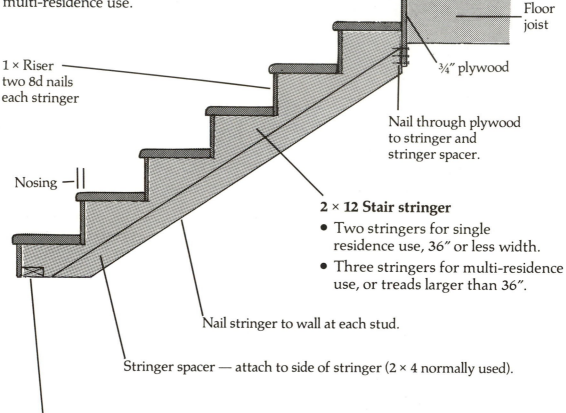

Subfloor sheathing

Floor joist

¾" plywood

1 × Riser two 8d nails each stringer

Nail through plywood to stringer and stringer spacer.

Nosing

2 × 12 Stair stringer
- Two stringers for single residence use, 36" or less width.
- Three stringers for multi-residence use, or treads larger than 36".

Nail stringer to wall at each stud.

Stringer spacer — attach to side of stringer (2 × 4 normally used).

If stringers are not attached to walls, notch stringers for bottom stop plate.

Chapter Seven

LAYOUT

Layout is the written language of the framer. If the lead framer "writes" clearly, then the framer reading the layout will be able to understand and properly perform the work. The layout person should include enough information in the layout so that the framer does not have to ask any questions. Layout language has been developed by framers over the years, and there are some variations. The version described in this manual is quite typical. Feel free to make any changes which reflect practices in your area. If you need to explain something about the layout that isn't shown in this chapter, either write it out on the plates in plain English, or explain it to the person who will do the framing.

Chapter ten provides a page with formulae for finding stud heights for different height foundations. This page may at first look a little complicated. However, after a closer look, you will find that by using the "Jobsite Worksheet" you can turn the difficult and often confusing job of finding stud heights into a fairly simple task.

WALL LAYOUT LANGUAGE

Layout

Studs
(cutaway, looking down in wall from above)

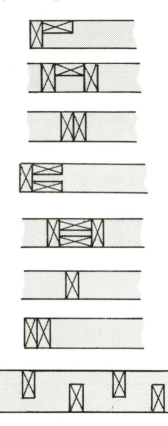

	Layout	
Corner	COR	
Backer	(X)	
Stud trimmer	X | T	
Double corner	D COR	
Double backer	DB	
Stud	X	
Double stud	X | X	
Staggered studs	ᵁ D ᵁ D — U = Up D = Down	
Beam	8/½T X|T|T — Write trimmer height (T) next to layout	
Nonbearing flat header	36" F	
Nonbearing L-header	39"L	
Nonbearing cripple header, more than 4'	60"C	
Bearing walls, solid header, sizes vary	39" (4x8)	
Header layout	X|T 34"(4x8) 35"c c|T|X / X|T|c c c c|T|X — C = Cripple	

WALL LAYOUT

Wall layout is the process of taking the information given on the plans and writing enough instructions, in layout language, on the top and bottom plates so the framer can build the wall without asking any questions.

There is some general information that must be considered before starting. These items are listed below. Unless otherwise noted, all layout discussions will assume 2 × 4 studs at 16" O.C. (on center).

Where possible, we want joists, studs, and rafters to set directly over each other.

Before layout is started, establish reference points in the building for measuring both directions of layout and use those points for joist, stud, and rafter/truss layout throughout the building. Check the building plans for special joist plan or rafter/truss plans indicating layout. Select a reference point which allows you to lay out in as long and straight a line as possible, and which ensures that a maximum number of rafters/trusses are directly supported by studs.

WALL LAYOUT STEPS

1) Spread top plate and bottom plate together in chalk lines. If plate is not long enough, cut top plate to break on middle of stud and four feet away from walls running into it.

2) Place plates in position with chalk lines.

3) Lay out for backers from chalk lines.

4) Lay out stud trimmers and cripples for windows and doors.

5) Lay out studs.

Chalking Lines

"Chalking lines" is the process of marking on the subfloor where the walls are to be placed.

Red chalk makes a permanent line and is easily seen. Blue chalk can be erased and is good to use if the lines might have to be changed. Using different colors allows you to distinguish between old and new lines.

Before chalking, when possible, check foundation and floor for square. Walls must be square, plumb, and level. If necessary, adjust your chalk lines accordingly.

Measurements for chalk lines are derived from wall dimensions as given in the plans. If the plans show finished walls, be sure to subtract the appropriate amount to get your framing measurements. Measure twice, chalk once.

LAYOUT WITH CHALK LINES

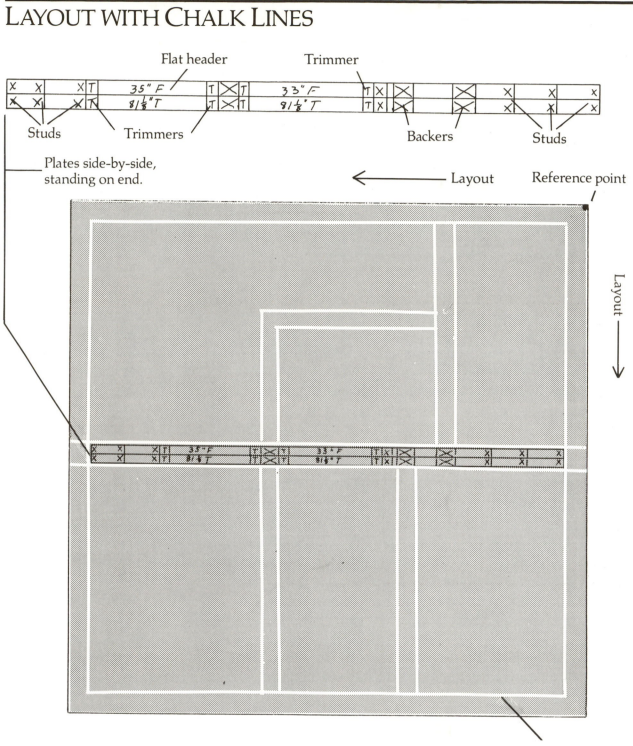

Chalk lines on subfloor sheathing using dimensions on plans.

PULLING LAYOUT FROM A CORNER (16" O.C.)

Pull layout so that sheathing will break in the middle of a joist/stud. Hook tape on the outside edge of rim joist/plate. Pull and locate 16" on tape, then measure back half the thickness of joist/stud (¾" for 2× stock) and mark. This puts the layout mark on 15¼". Make an "X" on the correct side of the layout mark to show the location of the joist/stud. Continue marking in this way for each subsequent 16" space, thus: 31¼", 47¼", 63¼", etc. Finish by squaring lines through the marks.

The reason for subtracting the ¾" is that the 4' × 8' sheathing will be installed from the *outside edge* of the rim joist/plate, not from the center.

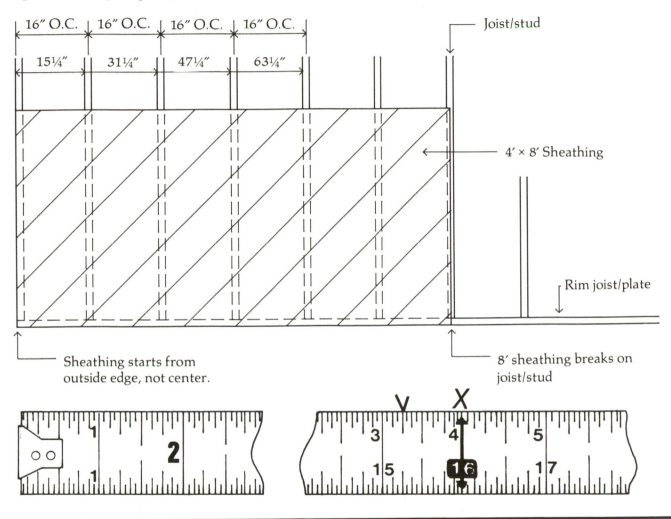

PULLING FROM AN EXISTING LAYOUT (16" O.C.)

Set end of tape on a layout mark, pull and mark 16", 32", 48", etc. Make an "X" on the correct side of the layout mark and square a line through the mark. For a long span, set a nail on the existing layout mark, hook tape on the nail, and pull layout.

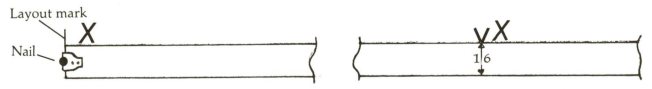

JOIST LAYOUT

Joist layout is the process of taking the information given on the plans and writing enough instructions on the top of the rim joist or double plate so the joist framer can spread and nail the joists without asking any questions.

Where possible, we want joists, studs, and rafters to set directly over each other.

Before layout is started, establish reference points in the building for measuring both directions of layout and use those points for joist, stud, and rafter/truss layout throughout the building. Check the building plans for special joist plan or rafter/truss plans indicating layout. Select a reference point which allows you to lay out in as long and straight a line as possible, and which ensures that a maximum number of rafters/trusses are directly supported by studs.

Check plans for openings in the floor required for stairs, chimneys, etc.

Check plans for bearing partitions on the floor. Double joists under bearing partitions running parallel.

Check locations of toilets to see if joists must be headed out for toilet drain pipe.

JOIST LAYOUT LANGUAGE

Joist	[____ ▨ X ▨]
Tail joist	[____ ▨ T ▨]
Double joist	[____ ▨ X│X ▨]
Beam	[____ ▨ 6×12 ▨]

JOIST LAYOUT STEPS

1) Layout for double joist, trimmer joist, and tail joist.
2) Layout for other joists.

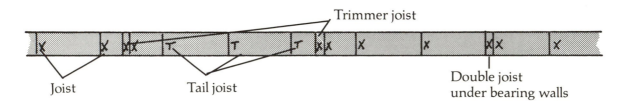

Trimmer joist

Joist   Tail joist   Double joist
                                    under bearing walls

ROOF LAYOUT

Roof layout is the process of taking the information given on the plans and writing enough instructions on the double plate for the roof framer to start on the roof.

Use the same reference points established for floor and wall building for starting layout on the roof.

Roof rafters and trusses are sometimes 24" O.C. as compared with 16" O.C. for floors and walls. In that case, only every third truss or rafter will be over a stud.

Before layout is started, check plans for openings in the roof required by dormers, skylights, chimneys, etc.

For roof trusses, lay out according to truss plan, especially for hip-truss packages.

ROOF LAYOUT LANGUAGE

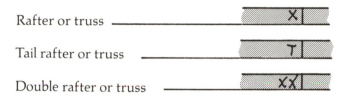

Rafter or truss

Tail rafter or truss

Double rafter or truss

ROOF LAYOUT STEPS

1) Layout for doubles, trimmers, and tail rafters or trusses.
2) Layout for other rafters or trusses.

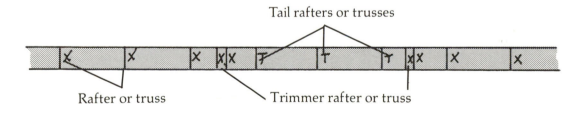

Tail rafters or trusses

Rafter or truss

Trimmer rafter or truss

Chapter Eight

MANAGING A FRAMING TEAM

Management techniques have developed over the years by studying and applying methods that work. The trend has been away from the dominating "command" approach and toward the cooperative "team" approach. This chapter deals with some organizational tasks, as well as with relationships and motivation. Developing good working relationships and instilling motivation is probably the most important and the most difficult task of a leader. A construction project manned by crews of skilled craftsmen who take pride in their work and get along with each other is bound to be successful. Assembling and directing such crews can only be accomplished by a leader who has developed good management skills.

Contents

LEAD FRAMERS

A Lead Framer Must Possess:

1) Knowledge to frame any building.

2) Ability to impart knowledge to other framers.

3) Ability to motivate other framers.

1. Knowledge to Frame Any Building

As a lead framer, you must thoroughly understand the basic concepts involved in framing any style building. The framing crew takes their direction from you; you, in turn, take your direction, depending on the situation, from any of the following:

a) Framing contractor

b) Site superintendent

c) Architect or engineer

d) Owner

There are a number of framing requirements that are easy to overlook. Compile a checklist of such items and refer to it during each phase of the job. For example:

Framing checklist

WALLS:	• Studs for tubs, medicine cabinets, etc.
	• Special shear nailing
	• Holddowns
	• Alignment of windows and doors
	• Size of studs
JOISTS:	• Heading-off for toilets
	• Double under bearing walls
	• Cantilevering overhangs
ROOF:	• Fascia overhang 6" for gutters
	• Skylights or roof hatch
	• Ventilation
	• Attic access
CRAWL SPACE:	• Vents in rim joists
	• Crawl-space access

2. Ability to Impart Knowledge to Other Framers

- When teaching someone, start with the basics. Assume nothing. Explain in clear and simple language exactly what the job is and how it is to be done.

- The easiest way to lead is to give orders, make demands, and threaten punishment. However, it creates an unsettling atmosphere which is not conducive to a cooperative, self-motivated crew. *Request* that framers do tasks; do not *order* them.

- Assume that no framer intentionally does something wrong. Do not be dogmatic or negative when you point out mistakes. Help him correct the error and show him how to avoid making it again.

- Treat each framer with respect. His time may be less valuable to the company, but his worth as an individual is equal to yours.

- The words, "please" and "thank-you," can make a framer feel much better about working for you. It is an easy way to let him know that what he does is important and appreciated.
- Do not give the hard, unpleasant jobs to the same framer time after time. The entire crew should share such tasks.
- When a framer asks you a question, give him the answer, but then explain how you got the answer so the next time he can figure it out himself.

3. Ability to Motivate Other Framers

To produce good work efficiently, a framer must be motivated.

To be motivated, a framer must:

A) Feel good about himself

B) Feel what he is doing is important

C) Be respected by his lead framer

D) Feel he is being treated equally

A) *Feel good about himself*

You are a lead framer, not a therapist, but your attitude toward your crew, as exemplified in items B, C, and D, should have a positive effect in this regard.

A crew whose members take pride in their individual and collective skills will invariably produce quality work and take pleasure in doing it.

B) *Feel what he is doing is important*

Every task, no matter how small, is necessary to complete the job and, therefore, important.

C) *Be respected by his lead framer*

1) Take time to listen and teach. If as lead framer, you are called upon to solve a framing problem, it is better to let the framer explain his solution first and, if it is an acceptable solution, let him do it his way. There are often several ways to solve a framing problem. If you have a way that is much faster or easier than the framer's way, explain it to him and tell him how you came to your conclusion.

2) Directions should be given in terms of the job, not the individual.

   For example:

   Negative — "I told you five minutes ago to build that wall."

   Positive — "We need that wall built right away so we can finish this unit."

3) Framers like to feel the person supervising is concerned about what they think and how they feel. Convey this through your words and actions.

D) *Feel he is being treated equally*

Don't show favoritism when assigning tasks. Make every effort to treat all framers fairly and take time to impartially deal with any complaints that might arise.

Goals for Motivation

Goals are excellent motivational tools.

When assigning each task, specify a goal for its completion within a certain amount of time.

Goals help you to:
1) Determine how fast you are getting work done.
2) Think of faster ways of getting work done.
3) Maintain the motivation to work faster.
4) Pass the day.

Goals should, of course, be realistic and attainable.

Develop three sets of goals:
1) Task goals (frame this wall in two hours).
2) Daily goals (frame and set six walls today).
3) Project goals (frame this building in three weeks).

Each framer should set his own daily goals. The lead framer should review them and, if the time assigned seems too long, should suggest ways to make the task easier or faster.

A framer's value to the company, and thus his wages, are commensurate with his ability to produce work that is high in both quantity and quality; if you produce more, you earn more. Making this fact clear to each member of your crew is, perhaps, the best motivation of all.

Company Goals
1) To provide income for framers and the company.
2) To provide a safe and enjoyable work environment.
3) To produce high-quality framed buildings.
4) To develop framing skills and abilities.
5) To build up the company.
6) To create comradeship among the framers.
7) To coordinate with the general contractor's schedule and needs.

It is great to have a team of framers working together and doing such a good and fast job that they feel good about themselves and the company they are a part of. This creates the desire to improve themselves and the company because they not only get positive feedback, but earn more money.

FRAMING CONTRACTOR TASKS

The framing contractor is, in many cases, also the lead framer. The following tasks are his responsibility.
1) Check location and quality of power supply.
2) Check location and date of lumber drop.
3) Check window delivery schedule.
4) Check truss delivery schedule when appropriate.
5) Arrange to have the builder complete as much site preparation as possible before starting, including leveling the area around the building where framers will be working.
6) Highlight items on plans that are easy to miss or hard to find.
7) Give the lead framer a list of potential problem areas and items that are easy to forget.

Chapter Nine

Costs, Equipment, and Material Handling

The aim of this chapter is to show how a number of organized management procedures can be integrated into the daily job routine. All these procedures have as their goal maximum productivity at minimal cost. Every member of a framing crew should be constantly aware of this goal and actively work to achieve it. For example, it is important that a framer be conscious of the value and cost of his time and of the materials he works with. When a framer drops a handful of nails, knowing the cost of those nails will help him decide whether it's worthwhile to break his flow and bend down and pick up the nails, or keep on working and forget about them. If a framer knows the actual time saved by a quick, as opposed to a relaxed walk, perhaps he will be convinced to pick up the pace.

The organization and handling of material is an often overlooked, and yet very important, aspect of framing productivity. It is amazing how much time and money can be saved by a little planning and applied common sense.

Contents

TIME / WAGE RELATION

Objective: To frame a quality building for the least possible cost. Wages represent the greatest percentage of the total cost, and since wages are paid for time spent working, it follows that if you lessen the time it takes to complete a building, you thereby reduce the wages paid and, thus, lower the total cost.

There are various tasks involved in framing. Some of the tasks can be done as fast or nearly as fast by the least experienced framer as by the most experienced framer, for example, carrying and spreading studs. Suppose that a $5/hour person can do the same job as a $10/hour person, only 20% slower. You save $4/hour by having the less expensive person do the job.

Be careful, however, because if an inexperienced framer takes on a job beyond his capabilities, the chances for mistakes are great. This could result in wages paid for work incorrectly done, and higher wages paid for an experienced framer to find and correct the mistake.

It is ideal for you to have a balanced crew of experienced and inexperienced framers. This allows you the flexibility to fit the framer to the task. It takes planning to coordinate framing tasks so that they are done as inexpensively as possible, but it is time well spent for the money you will save.

The following are job titles and responsibilities for a typical large framing crew, divided into four categories.

Organizing

1) **Lead Framer**
   - Coordinate work schedule with framing contractor.
   - Locate lumber drop.
   - Make sure tools and supplies are available.
   - Train framers.
   - Solve problems.

2) **Cut and Chalk Framer**
   - Cut stairs and rafters.
   - Chalk lines.
   - Assign framers to tasks.

3) **Layout Framer**
   - Layout for walls, joists, rafters, and trusses.

Walls

1) **Plumb and Line Framer**
   - Plumb and line walls.

2) **Wall Spreader**
   - Spread all parts of a wall (studs, trimmers, cripples, headers, etc.) so they are ready to be nailed together.

3) **Wall Nailer**
   - Nail walls together.

4) **Makeup Framer**
   - Nail together a quantity of stud-trimmers, corners, backers, headers, etc., before they are carried to the walls.

Floors

Joisting

1) **Rim Joister**
   - Nail joists in position.
   - Header out joists.

2) **Joist Nailer**
   - Nail joists in position.
   - Nail on joist hangers.
   - Nail blocking and drywall backing.

3) **Joist Spreader**
   - Carry joists into place.

TIME/WAGE RELATION *(continued)*

Sheathing
1) **Sheathing Setter**   – Place and set sheathing.
2) **Sheathing Nailer**   – Glue joists ahead of setter and nail sheathing behind setter.
3) **Sheathing Carrier**   – Carry sheathing to setter.

Roofs
1) **Spreaders**   – Spread and install trusses and rafters.
    – Install fascia.
2) **Blockers**   – Install blocking and drywall backing.
    – Set sheathing.
3) **Sheathing Packers**   – Carry sheathing to setters.

MULTIPLE FRAMER TASKS

Some tasks require more than one framer, for example, lifting large walls. Any time framers have to be called from other tasks to perform a common task, much care must be taken not to waste time. The cost per minute of a 5-framer task is 5 times that of a single framer task. The person organizing the task needs to take responsibility for making sure the task is ready for all framers. If it is lifting a wall, the organizing framer should make sure the wall is completely ready so framers don't stand around while one makes last-minute adjustments.

Learning Curve

Studies have been done which show that as output is doubled, the time required decreases according to a constant ratio. The common ratio is about 4 to 5, or 20%. For example, the fourth set of stairs built will take 20% less time than the second.

Multiple Cutting Analysis

Multiple cutting becomes efficient when you have to cut a number of pieces of lumber the same size. Trimmers, cripples, and blocks are good examples.

To multiple cut, first spread all the lumber to be cut out on the floor or a table. Then measure each piece, mark each piece with your square, and finally cut each piece.

Analysis has shown it takes 36% less time to cut ten pieces of 2 × 4 when the tasks of spreading, then measuring, then marking, and then cutting are done for all the pieces at one time.

For very large numbers of cuts it may be worthwhile to make a template or a measuring/cutting jig.

MOTION ANALYSIS

Question: Can significant time (and money) be saved by moving faster?

Motion analysis studies of framers produced the following results for different speeds over a given distance.

Relaxed walk . . . . 148 seconds
Quick walk . . . . 117 seconds
Nail pouch shuffle . . . . 84 seconds
Drop the nail pouch and run . . . . 63 seconds

Example: Let's suppose a framer walks 2 hours in an 8-hour day. If a framer does a quick walk instead of a relaxed walk, 25 minutes would be saved during the day. The nail pouch shuffle would save 52 minutes.

Conclusion: A quick pace pays.

SPEED VERSUS QUALITY

Speed or quality: Which should it be? How good must the work be considering it must be done as fast as possible? The two variables to consider when answering these questions are:

- Strength
- Attractiveness

First and most important is the structural integrity of the building. The second is to create a finished frame that will be pleasing to the eye.

Once requirements for strength and attractiveness are satisfied, the faster the job can be done, the better.

MATERIAL MOVEMENT

Framing requires a lot of material movement. It is estimated that one-quarter to one-third of a framer's time is spent moving material, so any time or energy saved is a cost reduction.

The following hints will help you save time, energy, motion, and in the last analysis, money.

- Whenever material is lifted or moved, it takes time and energy; therefore, move material as little as possible.
- When stacking lumber, consider the following:
  - Where will it be used next?
  - Will it be close to where it is going to be used?
  - Will it be in the way of another operation?
  - Will it obstruct a pathway?
- Always stack material neatly. This helps to keep the lumber straight and makes it easy for framers to pick up and carry it. Stack 2 × 4 studs in piles of eight for a convenient armful.
- Have second-floor lumber dumped close to the building so framers can stand on the lumber stack and throw it onto the second floor.
- When stacking lumber on a deck, place it where walls will not be built so it will not have to be moved again.
- Use mechanical aids, such as levers, for lifting. Remember your physics — the longer the handle in relation to the lifting arm, the easier it will be to lift the load.
- Two trips to the lumber stack or tool truck cost twice as much as one trip. If you have to go to the tool truck for a tool, check to see if you need nails or anything else.

TOOL MAINTENANCE SCHEDULE

Draw up a schedule such as this for your specific equipment; post it, and assign a reliable crew member to take charge of it.

Equipment	Schedule	Maintenance Operation	Lubrication
Worm-drive saw	Every Monday morning	Check oil	Heavy-duty saw lubricant
Nail guns	Each time before using	Oil	Gun oil
Compressors	Before using every day	Check oil	30-weight non-detergent
	Every Friday	Empty air and drain tanks	
	First of each month	Check air-intake filter	
	Every month	Change oil	30-weight non-detergent
Electric cords	First of each month	Test cords color code	

NAIL COSTS (approximate)

Nail	Cost per Box	Cost per Nail
16d	$15	½ cent
8d	$16	¼ cent
16d box galvanized	$25	1¼ cents
1½" roofing	$25	$1/3$ cent
2½" concrete	$1.20/lb.	$1 2/3$ cents
10d × .131 gun	$28	½ cent
8d gun	$28	½ cent
1½" roofing gun	$42	$2/3$ cent
1¾" gun staple	$50	½ cent
Shot and pin	—	38 cents

Chapter Ten

CODES, STANDARDS, AND SAFETY

Building code books are the ruling authorities when it comes to questions of framing. They are vast in scope and contain thousands of items which are of no use to the average framer. The code guides in this chapter make it easy for framers to locate in the various code books, the important things they need to know.

The "Standard Framing Dimensions" tables were developed to ensure the consistency and compatibility of all dimensions within a specific framing project. These tables can, of course, be adapted to any given stud height. When making your own table, be sure to double-check your calculations and review the dimensions with the builder. Mistakes are easier and less costly to eliminate before the job begins than to correct after they have been cut and nailed in place.

"Framing Terms" have all been defined by a framer for framers, and are as straightforward as possible.

FRAMER'S GUIDE TO THE 1994
UNIFORM BUILDING CODE (UBC)

Subject	Section	Table	Page
Beam and girders next to concrete	2317.6		313
Beam notches	2306.2.2		310
Crawl-space clearance and access	2317.3		312
Crawl-space ventilation	2317.7		313
Fire-stops and draft-stops	708		113 & 114
Joist bearing	2326.8.2 & .3		320
Joist notching and	2326.8.3		320
around openings	2326.8.4		320
Joist and rafters, bridging and	2306.7		310
blocking	2326.8.3		320
Joist under partitions	2326.8.5		320
Nailing patterns		23-I-Q	325 & 326
Nail penetration	2311.3.3	23-I-G	311
Nailing gypsum board wall sheathing		25G	352
Size, height and spacing of wood studs		23-1-R-3	64*
Nails	2311.3		311
Spans, ceiling joists		23-1-V-J-3 & 4	46-47*
		23-1-X-1 & 2	23-43*
Spans, floor joists		23-1-V-J-1 & 2	44-45*
		23-1-X-1 & 2	23-43*
Spans, plywood		23-1-S-1	328
		23-1-T-1	329
Spans, rafters		23-1-V-R-1 to 12	48 to 59*
		23-1-X-1 & 2	23 to 43*
Stairways	1006		181-184
Rafter ridge boards	2326.12.3		324
Treated wood	2317.4 & .5		313
Wall bracing	2326.11.3	23-I-W	322
			330
Wall double plates	2326.11.2		321
Wall foundation cripples	2326.11.5		323
Wall headers	2326.11.6		323

* These pages are found in "Dwelling Construction Under the Uniform Building Code." This is a shortened version of the U.B.C. that was written for the convenience of the home builder. It contains only 90 pages, as compared with the 1,050 pages of the U.B.C.

FRAMER'S GUIDE TO THE 1996
NATIONAL BUILDING CODE (BOCA)

Subject	Section	Table	Page
Beam and girders next to concrete	2305.6.3		257
Beam notches	2305.3		254-256
Attic access	1211.2		141
Crawl-space ventilation	1210.2		141
Fire-stops and draft-stops	721 and 806		81-82, 87
Joist bearing on girders	2305.6		257
Joist notching	2305.3.1		256
Joist and rafter bridging	2305.16		259
Nailing patterns		Table 2305.2	255-256
Nail gypsum board wall sheating		Table 2305.2	255-256
Nailing shear gypsum		Table 2305.2	255-256
Spans, plywood	2307.3	Tables 2307.3.3, 2307.3.1(1), 2307.3.1(2)	265-266
Stairways - interior	1014.11		120
Stairways - exterior	1014.12		121
Handrails	1014.6.6.1 1014.7 1022.0		119 119 128-129
Treated wood	2310.0 2311.0		267-268 268-269
Wall bracing	2305.4.2 2305.7 2305.8 2305.9	Table 2305.8.1	256-258
Wall double plates	2305.4.2		256-257
Wall headers	2305.11		259
Wood framing general	2303.0 2305.0 2306.0		253-264

FRAMER'S GUIDE TO THE 1994
STANDARD BUILDING CODE (SBCCI)

Subject	Section	Table	Page
Beams and girders next to concrete	2307.2		562
Crawl-space clearance	2304.3		554
Crawl-space access	1804.6.3.2		454
Crawl-space ventilation	1804.6.3.1		454
Fire-stops and draft-stops	705 and 2305.2		157-171 556-557
Joist bearing	2307.3.3		562
Joist notching	2307.3.6		563
Joist around openings	2307.4		563
Joist bridging	2307.3.4		562
Joist blocking	2307.3.5		563
Joist under partitions	2307.5		563
Nailing patterns		Table 2306.1	558-561
Nailing gypsum board, wall sheathing		Table 2306.1	559-560
Nailing shear plywood		Table 2308.1D	570
Spans, plywood floor-roof		Table 2307.6B	565
Spans, plywood walls		Table 2308.1B	568
Stairways	2307.9 1007		567 261-265
Rafter ridge boards	2309.1.2		587
Treated wood	2304		552-555
Wall bracing	2308.2.2	Table 2308.2.2A Table 2308.2.2B	571-572
Wall double plates	2308.1.4	Table 2308.3J	568 584-585
Wall headers	2308.3	Table 2308.3A-J	573-585
Attic access	2309.6		590
Attic ventilation	2309.7		590

FINDING STUD HEIGHTS FOR DIFFERENT HEIGHT FOUNDATIONS

SH = Stud Height
TRD = Transit Reading Down
TRU = Transit Reading Up
SB = Stud Base

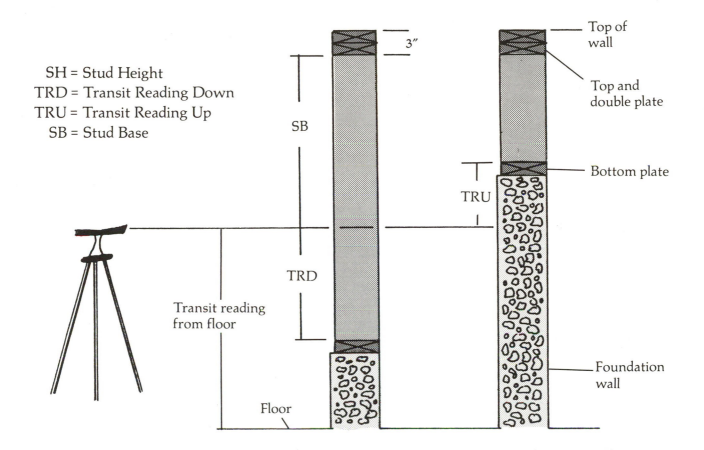

Equations

SH = TRD + SB

SH = SB – TRU

Steps

1) Set the bottom plate.
2) Find SB (stud base) by looking on the plans for height of wall; then subtract the height of the "Transit reading from floor" and the 3" for top and double plate.

Job Site Worksheet

	TRD	+	SB	=	SH
SH1		+		=	
SH2		+		=	
SH3		+		=	
SH4		+		=	
SH5		+		=	
SH6		+		=	
	SB	–	TRU	=	
SH7		–		=	
SH8		–		=	
SH9		–		=	
SH10		–		=	
SH11		–		=	
SH12		–		=	

STANDARD FRAMING DIMENSIONS
88⅝" STUDS

		Header Size	Trimmer Size* ** ***
Stud height . . . . 88⅝"			
Wall height . . . . 93⅛"			
R.O. windows . . . . Width–nominal			
	Height–nominal	4 × 8	81⅛"
		4 × 10	79⅛"
R.O. exterior doors . . . . . . Width–nominal + 2½"			
	Height–82⅝"	4 × 8	81⅛"
	Height–82⅛"	4 × 10	80⅝" cut T.P.
R.O. sliding glass doors . . . . Width – nominal			
	Height–6'10" door 82⅛"	4 × 8	81⅛"/½" furr
	82⅛"	4 × 10	80⅝" cut T.P.
	Height–6'8" door 80⅛"	4 × 8	81⅛"/2½" furr
	80⅛"	4 × 10	79⅛."/½" furr
R.O. interior doors . . . . . . Width–nominal + 2"			
(nonbearing)	Height–82⅝"		81⅛"
R.O. bifold doors . . . . Width–nominal + 1¼" for ½" drywall –nominal + 1½" for ⅝" drywall			
	Height – 82⅝"		81⅛"
R.O. bypass doors . . . . Width–nominal			
	Height–82⅝"		81⅛"
R.O. pocket doors . . . . Width–2 × nominal + 1"			
	Height–84½"		83"
Bathtubs . . . . Width–nominal + ¼"			
Tub fire blocks . . . . .14½" from finish floor to bottom of block			
Medicine-cabinet blocks . . . . . . .R.O. 14½" × 24"			
	Height–48" from finish floor to bottom of R.O. 3" minimum away from wall corner		

These dimensions should be checked with the job-site superintendent before beginning each job.

   *Furr = furring under header after header is in place.

  **Trimmer heights will increase by 1½" if lightweight concrete is used or ¾" if gypcrete is used.

 ***Cut T.P. – Cut the top plate out and leave the double plate.

   R.O. (rough opening) – Any opening framed by the framing members.

STANDARD FRAMING DIMENSIONS
92⅝" STUDS

		Header Size	Trimmer Size* **
Stud height . . . . 92⅝"			
Wall height . . . . 97⅛"			
R.O. windows . . . . Width–nominal			
	Height–nominal	4 × 8	85⅛"
		4 × 10	83⅛"
R.O. exterior doors . . . . . . Width–nominal + 2½"			
	Height–82⅝"	4 × 8	85⅛"/4" furr
	Height–82⅝"	4 × 10	83⅛"/2" furr
R.O. sliding **glass doors** . . . . Width–nominal			
	Height–6'10" door 82⅛"	4 × 8	85⅛"/4½" furr
	82⅛"	4 × 10	83⅛"/2½" furr
	Height–6'8" door 80⅛"	4 × 8	85⅛"/6½" furr
	80⅛"	4 × 10	83⅛"/4½" furr
R.O. interior doors . . . . . . Width–nominal + 2" (nonbearing)			
	Height–82⅝"		81⅛"
R.O. bifold doors . . . . Width–nominal + 1¼" for ½" drywall –nominal + 1½" for ⅝" drywall			
	Height–82⅝"		81⅛"
R.O. bypass doors . . . . Width–nominal			
	Height–82⅝"		81⅛"
R.O. pocket doors . . . . Width–2 × nominal + 1"			
	Height–84½"		83"
Bathtubs . . . . Width–nominal + ¼"			
Tub fire blocks . . . . 14½" from finish floor to bottom of block			
Medicine-cabinet **blocks** . . . . R.O. 14½" × 24"			
	Height–48" from finish floor to bottom of R.O. 3" minimum away from wall corner		

These dimensions should be checked with the job-site superintendent before beginning each job.

*Furr = furring under header after header is in place.

**Trimmer heights will increase by 1½" if lightweight concrete is used or ¾" if gypcrete is used.

R.O. (rough opening) – Any opening framed by the framing members.

STANDARD FRAMING DIMENSIONS

To adapt the standard framing dimensions to other stud heights, the following apply:

- All widths will be the same.
- The height for the windows will be nominal. To determine the trimmer size, just find the height of the bottom of the header and subtract 1½″ for the bottom plate.
- To determine the height of the window and, thereby, the amount of furring, you need to look on the plans. Usually the height can be found on the elevations. However, it can also appear on the floor plans. If the height is not given use the door height for the window height.
- For exterior doors, furr down to 82⅝″ from the subfloor sheathing.
- For sliding glass doors: – 6′10″ furr down to 82″
  – 6′8″ furr down to 80″.
- All else remains the same.

HEADER LENGTH FOR DOORS

Common Wood Doors

Door size	+ Jam shim allowance	+ Trimmer width	= Header size
Examples interior 2⁶ door (30")	+ 2"	+ 3"	= 35"
exterior 3⁰ door (36")	+ 2½"	+ 3"	= 41½"

HEADER LENGTH FOR WINDOWS

Aluminum or Vinyl Nail-Flange Windows

Window size	+ Trimmer width	= Header size
Examples 3⁰ (36")	+ 3"	= 39"
8⁰ (96")	+ 6"	= 102"

TRIMMER HEIGHT

Use the standard 81⅛" trimmer height for a standard 82⅝" opening.

The trimmer height must be adjusted for the following.

1) Special door sizes.
2) Special floor coverings that go on top of the subfloor and are applied before the doors are hung, such as lightweight concrete or gypcrete.

BATHROOM FRAMING DIMENSIONS

Where bathtub plumbing occurs, place studs at 8", 24", and approximately 32" from back wall (depending on width of tub) to allow space for piping.

Check medicine cabinets for spacing studs. Common medicine cabinet spacing is 14½" between studs and at least 3" from a corner. A standard opening is usually 24" from top to bottom and begins 48" from the floor.

Toilet Clearance

These are common dimensions, but should be checked with the builder.

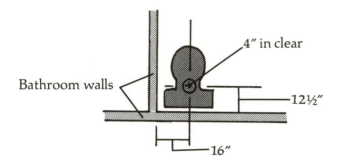

SAFETY TIPS

1) Keep work area neat underfoot, especially main pathways.

2) Don't leave nails sticking out; pull them or bend them over.

3) When lifting, always lift using your leg muscles and not your back. Always keep your back straight, your chin tucked in, and your stomach pulled in. Don't carry objects resting against your stomach; this will cause your spine to bend backward. Maintain the same posture when setting objects down. (See illustration.)

4) When working on joists, trusses, or rafters, always watch each step to see that what you are stepping on is secure.

5) Never pin circular saw guards back.

6) Do not wear loose or torn clothing that can get caught in tools.

7) Wear heavy shoes or boots that help protect your feet against injury.

8) Wear hard hats where material or tools can fall on your head.

9) Wear safety glasses where there is any potential for injury to the eye.

10) Use hearing protection when operating loud machinery or when hammering in a small, enclosed space.

11) Wear a dust mask for protection from sawdust, insulation fibers, or the like.

12) Use a respirator whenever working in exposure to toxic fumes.

GLOSSARY

Anchor bolt
A device for connecting wood members to concrete or masonry.

Backer
Three studs nailed together in a U-shape to which a partition is attached.

Bargeboard
Board attached to a gable rafter to which a rake board is attached.

Beam
A large horizontal structural member of wood or steel.

Bearing wall
A wall which supports the load of the structure above it.

Bevel
Any angle not at 90°. Also, a tool for marking such an angle.

Bird's-mouth
The notch in a rafter that rests on the top plate of a wall.

Box nails
Nail similar to a common nail, but having a smaller diameter shank.

Bridging
Bracing installed in an X-shape between floor joists to stiffen floor and distribute *live loads*. Also called cross-bridging.

Building line
The bottom measurement of a rafter's run, the top being the ridge line; the plumb cut of the bird's-mouth.

Camber
A slight crown or arch in a horizontal structural member.

Cantilever
Any part of a structure that projects beyond its main support and is balanced on it.

Cant strip
A length of lumber with a triangular cross-section used around edges of roofs or decks to help waterproof.

Casing
A piece of wood or metal trim that finishes off the frame of a door or window.

Casing nail
A large finish nail with a cone-shaped head.

Chalk line
A string covered with chalk used for marking straight lines.

Check
A lengthwise crack across the grain in a piece of lumber.

Chisel
A wedge-like, sharp-edged tool used for cutting or shaping wood.

Chord
Any principal member of a truss. In a roof truss the top chord replaces a rafter, and the bottom chord replaces a ceiling joist.

Collar beam
A horizontal board that connects pairs of rafters on opposite roof slopes.

Column
A vertical structural member.

Common nail
Nail used in framing and rough carpentry having a flat head about twice the diameter of its shank.

Corner
Two studs nailed together in an L-shape, used to attach two walls and provide drywall backing.

Cornice
The horizontal projection of a roof overhang at the *eaves*, consisting of lookout, soffit, and fascia.

Crawl space
The area bounded by foundation walls, first-floor joists, and the ground in a basementless house.

Cricket
A small, sloping structure built on a roof to divert water, usually away from a chimney. Also called a saddle.

Cripple
A short stud installed above or below a horizontal member in a wall opening.

Cripple jack rafter
A rafter which runs between a hip rafter and a valley rafter.

Crowbar
A steel bar usually slightly bent at one or both ends, used as a lever.

Crown
The high point of a piece of lumber with a curve in it.

Cup
Warp across the grain.

d
Abbreviation for penny. The abbreviation comes from the Roman word "denarius," meaning coin, which the English adapted to penny. It originally referred to the cost of a specific nail per 100. Today it refers only to nail size.

Dead load
The weight of all structure in place.

Dormer
A structure with its own roof projecting from a sloping roof.

Double cheek cut
A two-sided cut that forms a V at the end of some rafters, especially in hip and gambrel roofs.

Eave
The part of a roof that projects beyond its supporting walls.

Face nailing
Nailing at right angles to the surface.

Fascia
A vertical board nailed to the lower ends of rafters that form part of a cornice.

Fire block
A short piece of framing lumber nailed horizontally between joists or studs to partially block the flow of air and, thus, to slow the spread of fire.

Footings
The base, usually poured concrete, on which the foundation wall is built. The footing and foundation wall are often formed and poured as a single unit.

Form
A mold of metal or wood used to shape concrete until it has set.

Foundation
The building's structural support below the first-floor construction. It rests on the footing, and transfers the weight of the building to the soil.

Frame
The wood skeleton of a building. Also called framing.

Framer
A person who does rough carpentry.

Framing anchor
A metal device for connecting wood framing members that meet at right angles.

Furring
Strips of wood fastened across studs or joists to a level or plumb nailing surface for finish wall or ceiling material, usually sheetrock. Also called strapping.

Gable
The triangular part of an end wall between the eaves and ridge of a house with a peaked roof.

Gable roof
A roof shape characterized by two sections of roof of constant slope that meet at a ridge; peaked roof.

Gambrel roof
A roof shape similar to a gable roof, but with two sections of roof on each side of the ridge, the lower section being steeper than the upper.

Girder
A main horizontal beam of steel or wood.

Grade
1) A designation of quality, especially of lumber and plywood.
2) Ground level. Also the slope of the ground on a building site.

Grain
The direction of fibers in wood.

Gusset
A flat piece of plywood or metal attached to each side of two framing members to tie them together, or strengthen a joint.

Hammer
A tool consisting of a metal head set perpendicular on a handle, used for driving nails.

Header
Any structural wood member used across the ends of an opening to support the cut ends of shortened framing members in a floor, wall, or roof.

Hip
The outside angle where two adjacent sections of roof meet at a diagonal. The opposite of a valley.

Hip rafter
The diagonal rafter which forms a hip.

Hip roof
A roof shape characterized by four or more sections of constant slope, all of which run from a uniform eave height to the ridge.

Jack rafter
A short rafter, usually running between a top plate and a hip rafter or between a ridge and a valley rafter.

Jack stud
A shortened stud supporting the header above a door or window. Also called a trimmer or jamb stud.

Jamb
The side of a window or door opening.

Joint
The line along which two pieces of material meet.

Joist
One of a parallel series of structural members used for supporting a floor or ceiling. Joists are supported by walls, beams, or girders.

Joist hanger
A metal framing anchor for holding joists in position against a rim joist, header or beam.

Kerf
The cut made by a saw blade.

Kneewall
A short wall under a slope, usually in attic space.

Lead framer
Foreman or leader of a crew.

Ledger
A strip of lumber attached to the side of a girder near its bottom edge to support joists. Also, any similar supporting strip.

Level
Perfectly horizontal.

Lightweight plate
A plate that goes under the bottom plate to raise walls for lightweight concrete or gypcrete.

Live load
The total variable weight on a structure. It includes the weights of people, furnishings, snow, and wind.

Lookout
A horizontal framing member between a stud wall and the lower end of a rafter to which the soffit is attached.

Lumber
Wood cut at a sawmill into usable form.

Makeup
Parts of a wall, such as backers, corners, headers, and stud trimmers, which are cut before the wall is spread.

Mansard roof
A type of roof with two slopes on each of four sides, the lower slope much steeper than the upper and ending at a constant eave height.

Mortise
A recess cut into wood.

Mudsill
The lowest plate in a frame wall. It rests on the foundation or slab.

Nail gun
A hand-operated tool powered by compressed air which drives nails.

Nominal size

The rounded-off, simplified dimensional name given to lumber; e.g., a piece of lumber whose actual size is 1½″ × 3½″ is given the more convenient, nominal designation of 2″ × 4″.

Nonbearing

A dividing wall that supports none of the structure above it.

Nosing

The rounded front edge of a stair tread that extends over the riser.

On center (O.C.)

Layout spacing designation which refers to distance from the center of one framing member to another.

Overhang

The part of a roof that extends beyond supporting walls.

Parallel

Extending in the same direction equidistant at all points.

Parapet

A low wall or rail at the edge of a balcony or roof.

Partition

An interior wall that divides a building into rooms or areas.

Party wall

A wall between two adjoining living quarters in a multi-family dwelling.

Penny

Word applied to nails to indicate size; abbreviated as "*d*".

Perpendicular

At right angles to a plane, or flat surface.

Pitch

The slope of a roof.

Plate

A horizontal framing member laid flat.

> *Bottom plate*
>
> The lowest plate in a wall in the platform framing system, resting on the subfloor, to which the lower ends of studs are nailed.
>
> *Double plate*
>
> The uppermost plate in a frame wall that has two plates at the top. Also called a cap or rafter plate.
>
> *Sill plate*
>
> The structural member, attached to the top of the foundation, that supports the floor structure. Also called a sill or mudsill.
>
> *Top plate*
>
> The framing member nailed across the upper ends of studs and beneath the double plate.

Platform framing

A method of construction in which wall framing is built on and attached to a finished box sill. Joists and studs are not fastened together as in balloon and braced framing.

Plumb

Straight up and down, perfectly vertical.

Plumb and line

The process of straightening all the walls so they are vertical and straight from end to end.

Plumb cut

Any cut in a piece of lumber, such as at the upper end of a common rafter, that will be plumb when the piece is in its final position.

Purlin

The horizontal framing members in a gambrel roof between upper and lower rafters.

Rabbet

A groove cut in or near the edge of a piece of lumber to receive the edge of another piece.

Rafter

One of a series of main structural members which forms a roof.

Rake

The finish wood member running parallel to the roof slope at the gable end.

Reveal

The surface left exposed when one board is fastened over another; the edge of the upper set slightly back from the edge of the lower.

Ridge

The horizontal board to which the top ends of rafters are attached.

Rim joist

Joist which forms the perimeter of a floor framing system.

Rise

1) In a roof, the vertical distance between the top of the double plate and the point where a line, drawn through the edge of the double plate and parallel to the roof's slope, intersects the center line of the ridgeboard.
2) In a stairway, the vertical height of the entire stairway measured from floor to floor.

Riser

The vertical board between two stair treads.

Roof sheathing

Material, usually plywood, laid flat on roof trusses or rafters to form the roof.

Rough opening (R.O.)

Any opening formed by the framing members to accommodate doors or windows.

Rout

To cut out by gouging.

Run

1) In a roof with a ridge, the horizontal distance between the edge of the rafter plate (building line) and the center line of the ridgeboard.
2) In a stairway, the horizontal distance between the top and bottom risers plus the width of one tread.

Scaffold

Any temporary working platform and the structure to support it.

Scribe

To mark for an irregular cut.

Seat cut

The horizontal cut in a bird's-mouth that rests on the double plate.

Shed roof

A roof that slopes in only one direction.

Shim

A thin piece of material, often tapered (such as a wood shingle) inserted between building materials for the purpose of straightening or making their surfaces flush at a joint.

Sill

1) A sill plate.
2) The structural member forming the bottom of a rough opening for a door or window. Also, the bottom member of a door or window frame.

Single cheek cut

A bevel cut at the end of a rafter, especially in hip and gambrel roofs.

Sleeper

Lumber laid on a concrete floor as a nailing base for wood flooring.

Slope

The pitch of a roof, expressed as inches of rise per 12″ of run.

Soffit

The underside of a projection, such as a cornice.

Solid bridging

Blocking between joists cut from the same lumber as the joists themselves and used to stiffen the floor.

Spacer

Any piece of material used to maintain a permanent space between two members.

Span

The distance between structural supports, measured horizontally.

Speed square

A triangle-shaped tool used for marking perpendicular and angled lines.

Square

1) At 90° or a right angle.
2) The process of marking and cutting at a right angle.
3) Any of several tools for marking at right angles and for laying out structural members for cutting or positioning.
4) A measure of roofing and some siding materials equal to 100 square feet of coverage.

Stair

A single step.

Stair nuts

Two screw clamps that are attached to a framing square for marking stair stringers.

Stairway

A flight of stairs, made up of stringers, risers, and treads.

Stairwell

The opening in a floor for a stairway.

Stickers

Strips of scrap wood used to create an air space between layers of lumber.

Stop

In general, any device or member that prevents movement.

Story pole

A length of wood marked off and used for repetitive layout or to accurately transfer measurements.

Stringer

In stairway construction, the diagonal member that supports treads and to which risers are attached.

Structural

Adjective generally synonymous with "framing."

Stud
The main vertical framing member in a wall to which finish material or other covering is attached.

Subfloor sheathing
The rough floor, usually plywood, laid across floor joists and under finish flooring.

Tail
The part of a rafter which extends beyond the double plate.

Tail joist
A shortened joist that butts against a header.

Tape
A measure of coiled flexible steel.

Template
A full-sized pattern.

Threshold
The member at the bottom of a door between the jambs.

Toenailing
To drive a nail at an angle to join two pieces of wood.

Tongue
The shorter and narrower of the two legs of a framing square.

Tool pouch
Pouch worn around the waist for holding tools and nails.

Tread
The horizontal platform of a stair.

Trimmer
The structural member on the side of a framed rough opening to narrow or stiffen the opening. Also the shortened stud (jack stud) which supports a header in a door or window opening.

Truss
An assembly for bridging a broad span, most commonly used in roof construction.

Utility knife
A hand-held knife with a razor-like blade, commonly used to cut drywall, sharpen pencils, etc.

Valley
The inside angle where two adjacent sections of a roof meet at a diagonal. The opposite of a hip.

Valley rafter
A diagonal rafter which forms a valley.

Wall puller
A tool used for aligning walls.

Wall sheathing
Material, usually plywood, attached to studs to form the outside wall and provide structural strength.

Wane
A defect in lumber caused at the mill by sawing too close to the outside edge of a log and leaving an edge either incomplete or covered with bark.

Warp
Any variation from straight in a piece of lumber; bow, cup, crook, twist.

Western framing
Platform framing.

Worm-drive saw
A circular power saw turned by a worm-gear drive. It is somewhat heavier and produces more torque on the blade than a standard circular saw.

FOR READING AND REFERENCE

Anderson, L. O., *Wood-Frame House Construction*. Revised edition. Washington, D.C.: U.S. Department of Agriculture (Handbook No. 73), 1970.

Badzinski, Stanley, Jr., *Roof Framing*. Englewood Cliffs, NJ: Prentice Hall, Inc., 1976.

Ball, John E., *Audel's Carpenters and Builders Library*. Four volumes. Indianapolis, IN: Audel/Sams, 1977.

Blackburn, Graham, *Illustrated Basic Carpentry*. Boston: Little, Brown, 1977.

Feirer, John and Hutchings, Gilbert, *Carpentry and Building Construction*. Peoria, IL: Chas. A. Bennett Co., Inc., 1981.

Fine Homebuilding Builder's Library, *Frame Carpentry*. Newton, CT: Tauton Publications, 1988.

Koel, Leonard, *Carpentry*. Homewood, IL: American Technical Publishers, Inc., 1985.

Locke, Jim, *The Apple Corps Guide to the Well-Built House*. Boston: Houghton Mifflin, 1988.

Maguire, Byron W., *Carpentry for Residential Construction*, Carlsbad, CA: Craftsman Book Company, 1987.

Maguire, Byron W., *Carpentry Framing and Finishing*, second edition. Englewood Cliffs, NJ: Prentice Hall, 1989.

National Lumber Manufacturers Association, *Manual for House Framing*. Washington, D.C.: National Forest Products Association, 1961.

Reed, Mortimer, *Residential Carpentry*. New York: John Wiley and Sons, Inc., 1980.

Sunset books, *Basic Carpentry Illustrated*. Menlo Park, CA: Lane Publishing Co.

Syvanen, Bob, *Carpentry: Some Tricks of the Trade from an Old-Style Carpenter*. Charlotte, NC: East Woods/Fast and McMillan, 1982.

Tetrault, Jeanne, *The Women's Carpentry Book*. New York: Anchor/Doubleday, 1980.

Tood, Ken, *Carpentry Layout*. Carlsbad, CA: Craftsman Book Co., 1988.

Wagner, Willis, *Modern Carpentry*. S. Holland, IL: Goodheart-Willcox, 1976.

Wahlfeldt, Bette, *Woodframe House Building: An Illustrated Guide*. Blue Ridge Summit, PA: Tab, 1988.

Wilson, J. Douglas, *Practical House Carpentry*. New York: McGraw-Hill, 1973.

INDEX

123

NOTES

NOTES

NOTES

NOTES

NOTES